W9-AHP-921

501
Math Word Problems

501

Math Word Problems

2nd Edition

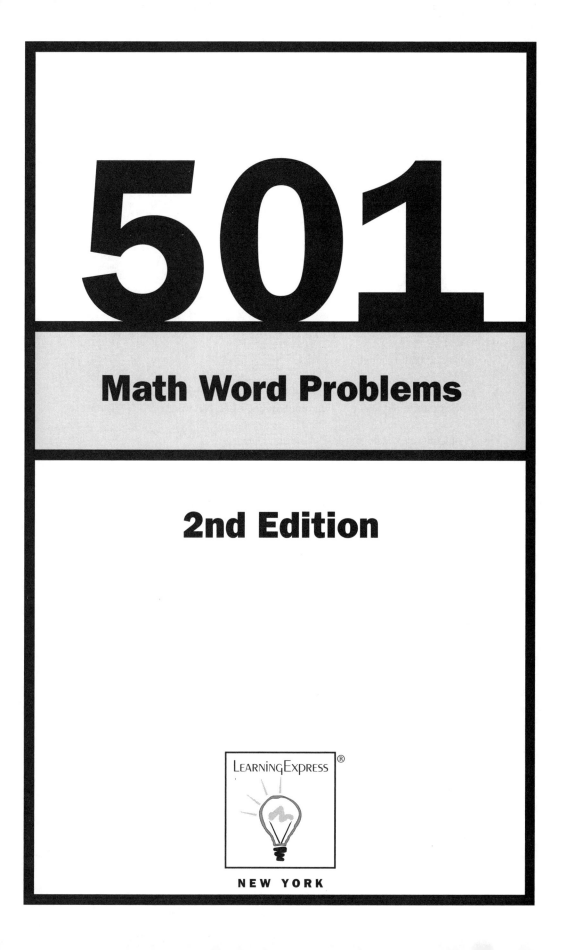

LEARNINGEXPRESS®

NEW YORK

Library of Congress Cataloging-in-Publication Data:
501 math word problems.—2nd ed.
    p. cm.
  ISBN 1-57685-563-5
  1. Mathematics—Problems, exercises, etc. 2. Word problems (Mathematics) I.
LearningExpress (Organization) II. Title: Five hundred one math word problems. III.
Title: Five hundred and one math word problems.
  QA43.A114 2006
  510.76—dc22

                                                                            2006045108

Printed in the United States of America

9 8 7 6 5 4 3 2 1

Second Edition

ISBN 1-57685-563-5

For more information or to place an order, contact LearningExpress at:
    55 Broadway
    8th Floor
    New York, NY 10006

Or visit us at:
    www.learnatest.com

Contents

Introduction

Welcome to *501 Math Word Problems!* This book is designed to provide you with review and practice for math success. It provides 501 problems so you can flex your muscles with a variety of mathematical concepts. *501 Math Word Problems* is designed for many audiences. It is for anyone who has ever taken a math course and needs to refresh and revive forgotten skills. It can be used to supplement current instruction in a math class. Or, it can be used by teachers and tutors who need to reinforce student skills. If at some point you feel you need further explanation about some of the more advanced math topics highlighted in this book, you can find them in other LearningExpress publications. *Algebra Success in 20 Minutes a Day, 501 Algebra Questions, Geometry Success in 20 Minutes a Day*, and *501 Geometry Questions* can help you with these complex math skills.

How to Use This Book

First, look at the table of contents to see the types of math topics covered in this book. The book is organized in six sections: Miscellaneous Math, Fractions, Decimals, Percents, Algebra, and Geometry. The structure follows a common sequence of math concepts. You may want to follow the sequence because the concepts grow more advanced as the book progresses. However, if your skills are just rusty, or if you are using this book to supplement topics you are currently learning, you may want to jump around from topic to topic.

As you complete the math problems in this book, you will undoubtedly want to check your answers against the answer explanation section at the end of each chapter. Every problem in *501 Math Word Problems* has a complete answer explanation. For problems that require more than one step, a thorough step-by-step explanation is provided. This will help you understand the problem-solving process. The purpose of drill and skill practice is to make you proficient at solving problems. Like an athlete preparing for the next season or a musician warming up for a concert, you become skilled with practice. If, after completing all the problems in a section, you feel you need more practice, do the problems again. It's not the answer that matters most—it's the process and the reasoning skills that you want to master.

You will probably want to have a calculator handy as you work through some of the sections. It's always a good idea to use it to check your calculations. If you have difficulty factoring numbers, the multiplication chart on the next page may help you. If you are unfamiliar with prime numbers, use the list on the next page so you won't waste time trying to factor numbers that can't be factored. And don't forget to keep lots of scrap paper on hand.

Make a Commitment

Success does not come without effort. Make the commitment to improve your math skills. Work for understanding. *Why* you do a math operation is as important as *how* you do it. If you truly want to be successful, make a commitment to spend the time you need to do a good job. You can do it! When you achieve math success, you have laid the foundation for future challenges and success. So sharpen your pencil and practice!

Multiplication Table

×	2	3	4	5	6	7	8	9	10	11	12
2	4	6	8	10	12	14	16	18	20	22	24
3	6	9	12	15	18	21	24	27	30	33	36
4	8	12	16	20	24	28	32	36	40	44	48
5	10	15	20	25	30	35	40	45	50	55	60
6	12	18	24	30	36	42	48	54	60	66	72
7	14	21	28	35	42	49	56	63	70	77	84
8	16	24	32	40	48	56	64	72	80	88	96
9	18	27	36	45	54	63	72	81	90	99	108
10	20	30	40	50	60	70	80	90	100	110	120
11	22	33	44	55	66	77	88	99	110	121	132
12	24	36	48	60	72	84	96	108	120	132	144

Prime Numbers < 1,015

2	3	5	7	11	13	17	19	23	29
31	37	41	43	47	53	59	61	67	71
73	79	83	89	97	101	103	107	109	113
127	131	137	139	149	151	157	163	167	173
179	181	191	193	197	199	211	223	227	229
233	239	241	251	257	263	269	271	277	281
283	293	307	311	313	317	331	337	347	349
353	359	367	373	379	383	389	397	401	409
419	421	431	433	439	443	449	457	461	463
467	479	487	491	499	503	509	521	523	541
547	557	563	569	571	577	587	593	599	601
607	613	617	619	631	641	643	647	653	659
661	673	677	683	691	701	709	719	727	733
739	743	751	757	761	769	773	787	797	809
811	821	823	827	829	839	853	857	859	863
877	881	883	887	907	911	919	929	937	941
947	953	967	971	977	983	991	997	1,009	1,013

501
Math Word Problems

Miscellaneous Math

This chapter consists of 63 problems dealing with basic math concepts including whole numbers, negative numbers, exponents, and square roots. It will provide a warm-up session before you move on to more difficult problems.

1. Bonnie has twice as many cousins as Robert. George has 5 cousins, which is 9 less than Bonnie has. How many cousins does Robert have?
   a. 17
   b. 22
   c. 4
   d. 7

2. Oscar sold 2 glasses of milk for every 5 sodas he sold. If he sold 10 glasses of milk, how many sodas did he sell?
   a. 45
   b. 20
   c. 25
   d. 10

3. Justin earned scores of 85, 92, and 95 on his science tests. What does he need to earn on his next science test to have an average (arithmetic mean) of 93%?

   a. 93

   b. 100

   c. 85

   d. 96

4. Brad's class collected 270 cans of food. They boxed them in boxes of 30 cans each. How many boxes did they need?

   a. 240

   b. 10

   c. 9

   d. 5

5. Joey participated in a dance-a-thon. His team started dancing at 10 A.M. on Friday and stopped at 6 P.M. on Saturday. How many hours did Joey's team dance?

   a. 52

   b. 56

   c. 30

   d. 32

6. Which expression has an answer of 18?

   a. $2 \times 5 + 4$

   b. $2 \times (4 + 5)$

   c. $5 \times (2 + 4)$

   d. $4 \times 2 + 5$

7. Callie's grandmother pledged $0.50 for every mile Callie walked in her walk-a-thon. Callie walked 9 miles. How much does her grandmother owe?

   a. $4.50

   b. $18.00

   c. $5.00

   d. $9.00

8. What is the square root of 64?

   a. 16

   b. 128

   c. 32

   d. 8

9. Mr. Brown plowed 6 acres in 1 hour. At this rate, how long will it take him to plow 21 acres?
   a. 3 hours
   b. 4 hours
   c. 3.5 hours
   d. 4.75 hours

10. What is the prime factorization of 84?
    a. 42×2
    b. $2 \times 2 \times 4 \times 6$
    c. $2 \times 7 \times 6$
    d. $2 \times 2 \times 3 \times 7$

11. What is 2^5?
    a. 10
    b. 64
    c. 32
    d. 16

12. The low temperature in Anchorage, Alaska today was $-8°F$. The low temperature in Los Angeles, California was $63°F$. What is the difference in the two low temperatures?
    a. 55°
    b. 71°
    c. 61°
    d. 14°

13. The Robin's Nest Nursing Home had a fundraising goal of $9,500. By the end of the fundraiser, they had exceeded their goal by $2,100. How much did they raise?
    a. $7,400
    b. $13,600
    c. $10,600
    d. $11,600

14. Mount Everest is 29,028 ft high. Mount Kilimanjaro is 19,340 ft high. How much taller is Mount Everest?
    a. 9,688 feet
    b. 10,328 feet
    c. 11,347 feet
    d. 6,288 feet

15. The area of a square is 36 cm². What is the length of one side of the square?

    a. 6 cm

    b. 9 cm

    c. 18 cm

    d. 24 cm

16. Mrs. Farrell's class has 26 students. Only 21 were present on Monday. How many were absent?

    a. 15

    b. 5

    c. 4

    d. 16

17. Lucy's youth group raised $1,569 for charity. They decided to split the money evenly among 3 charities. How much will each charity receive?

    a. $784.50

    b. $423.00

    c. $523.00

    d. $341.00

18. Jason made 9 two-point baskets and 3 three-point baskets in Friday's basketball game. He did not score any other points. How many points did he score?

    a. 21

    b. 12

    c. 24

    d. 27

19. Jeff left Hartford at 2:15 P.M. and arrived in Boston at 4:45 P.M. How long did the drive take him?

    a. 2.5 hours

    b. 2.3 hours

    c. 3.25 hours

    d. 2.75 hours

20. Shane rolls a die numbered 1 through 6. What is the probability Shane rolls a 3?

  a. $\frac{3}{6}$

  b. $\frac{1}{6}$

  c. $\frac{1}{3}$

  d. $\frac{1}{2}$

21. Susan traveled 114 miles in 2 hours. If she keeps going at the same rate, how long will it take her to go the remaining 285 miles of her trip?

  a. 5 hours

  b. 3 hours

  c. 7 hours

  d. 4 hours

22. A flight from Pittsburgh to Los Angeles took 5 hours and covered 3,060 miles. What was the plane's average speed?

  a. 545 mph

  b. 615 mph

  c. 515 mph

  d. 612 mph

23. Larry purchased 3 pairs of pants for $18 each and 5 shirts for $24 each. How much did Larry spend?

  a. $42

  b. $54

  c. $174

  d. $186

24. How many square centimeters are in one square meter?

  a. 100 sq cm

  b. 10,000 sq cm

  c. 144 sq cm

  d. 100,000 sq cm

25. Raul's bedroom is 4 yards long. How many inches long is the bedroom?

  a. 144 inches

  b. 48 inches

  c. 400 inches

  d. 4,000 inches

26. Jeff burns 500 calories per hour bicycling. How long will he have to ride to burn 1,250 calories?

    a. 3 hours

    b. 2.5 hours

    c. 1.5 hours

    d. 2 hours

27. The temperature at 6 P.M. was 31°F. By midnight, it had dropped 40°F. What was the temperature at midnight?

    a. 9°F

    b. −9°F

    c. −11°F

    d. 0°F

28. The total ticket sales for a soccer game were $1,260; 210 tickets were purchased. If all the tickets are the same price, what was the cost of a ticket?

    a. $6.00

    b. $3.50

    c. $10.00

    d. $7.50

29. Sherman took his pulse for 10 seconds and counted 11 beats. What is Sherman's pulse rate in beats per minute?

    a. 210 beats per minute

    b. 110 beats per minute

    c. 66 beats per minute

    d. 84 beats per minute

30. Jennifer flipped a coin four times and got heads each time. What is the probability that she gets heads on the next flip?

    a. 1

    b. $\frac{1}{32}$

    c. $\frac{1}{2}$

    d. 0

31. Jody's English quiz scores are 56, 93, 72, 89, and 87. What is the median of her scores?

    a. 72

    b. 87

    c. 56

    d. 85.6

32. What is the greatest common factor of 24 and 64?

    a. 8

    b. 4

    c. 12

    d. 36

33. Twelve coworkers go out for lunch together and order three pizzas. Each pizza is cut into eight slices. If each person gets the same number of slices, how many slices will each person get?

    a. 4

    b. 3

    c. 5

    d. 2

34. Marvin is helping his teachers plan a field trip. There are 136 people going on the field trip and each school bus holds 51 people. What is the minimum number of school buses they will need to reserve for the trip?

    a. 3

    b. 2

    c. 4

    d. 5

35. Which number in the answer choices below is not equivalent to the other numbers?

    a. 0.6

    b. 60%

    c. $\frac{3}{5}$

    d. 6%

36. Lance has 70 cents, Margaret has three-fourths of a dollar, Guy has two quarters and a dime, and Bill has six dimes. Who has the most money?
  a. Lance
  b. Margaret
  c. Guy
  d. Bill

37. The students at Norton School were asked to name their favorite type of pet. Of the 430 students surveyed, 258 said that their favorite type of pet was a dog. Suppose that only 100 students were surveyed, with similar results, about how many students would say that a dog is their favorite type of pet?
  a. 58
  b. 60
  c. 72
  d. 46

38. A group of five friends went out to lunch. The total bill for the lunch was $53.75. Their meals all cost about the same, so they wanted to split the bill evenly. Without considering tip, how much should each friend pay?
  a. $11.25
  b. $12.85
  c. $10.75
  d. $11.50

39. The value of a computer is depreciated evenly over five years for tax purposes (meaning that every year the computer is worth less by the same amount and that it is worth $0 after five years). If a business paid $2,400 for a computer, how much will it have depreciated after 2 years?
  a. $480
  b. $1,200
  c. $820
  d. $960

40. Steve earned a 96% on his first math test, a 74% on his second test, and an 85% on his third test. What is his test average?
  a. 91%
  b. 85%
  c. 87%
  d. 82%

41. A national park keeps track of how many people per car enter the park. Today, 57 cars had 4 people, 61 cars had 2 people, 9 cars had 1 person, and 5 cars had 5 people. What is the average number of people per car? Round to the nearest person.

    a. 2

    b. 3

    c. 4

    d. 5

42. A large pipe dispenses 750 gallons of water in 50 seconds. At this rate, how long will it take to dispense 330 gallons?

    a. 14 seconds

    b. 33 seconds

    c. 22 seconds

    d. 27 seconds

43. The light on a lighthouse blinks 45 times a minute. How long will it take the light to blink 405 times?

    a. 11 minutes

    b. 4 minutes

    c. 9 minutes

    d. 6 minutes

44. A die is rolled and a coin is tossed. What is the probability that a 3 will be rolled and a tail tossed?

    a. $\frac{1}{2}$

    b. $\frac{1}{6}$

    c. $\frac{1}{12}$

    d. $\frac{1}{8}$

45. Wendy has 5 pairs of pants and 7 shirts. How many different outfits can she make with these items (each outfit consists of one pair of pants and one shirt)?

    a. 12

    b. 24

    c. 35

    d. 21

46. Audrey measured the width of her dining room in inches. It is 150 inches. How many feet wide is her dining room?

   a. 12 feet

   b. 9 feet

   c. 12.5 feet

   d. 10.5 feet

47. Sharon wants to make 15 half-cup servings of soup. How many ounces of soup does she need?

   a. 60 ounces

   b. 150 ounces

   c. 120 ounces

   d. 7.5 ounces

48. Justin weighed 8 lb 12 oz when he was born. At his two-week check-up, he had gained 8 ounces. What was his weight in pounds and ounces?

   a. 9 lb

   b. 8 lb 15 oz

   c. 9 lb 4 oz

   d. 10 lb 2 oz

49. One inch equals 2.54 centimeters. The dimensions of a table made in Europe are 85 cm wide by 120 cm long. What is the width of the table in inches? Round to the nearest tenth of an inch.

   a. 30 inches

   b. 215.9 inches

   c. 33.5 inches

   d. 47.2 inches

50. A bag contains 3 red, 6 blue, 5 purple, and 2 orange marbles. One marble is selected at random. What is the probability that the marble chosen is blue?

   a. $\frac{4}{13}$

   b. $\frac{3}{8}$

   c. $\frac{3}{16}$

   d. $\frac{3}{5}$

51. The operator of an amusement park game kept track of how many tries it took participants to win the game. The following is the data from the first ten people:

2, 6, 3, 4, 6, 2, 8, 4, 3, 5

What is the median number of tries it took these participants to win the game?

a. 8

b. 6

c. 4

d. 2

52. Max goes to the gym every fourth day. Ellen's exercise routine is to go every third day. Today is Monday and both Max and Ellen are at the gym. What will the day of the week be the next time they are BOTH at the gym?

a. Sunday

b. Wednesday

c. Friday

d. Saturday

53. Danny is a contestant on a TV game show. If he gets a question right, the points for that question are added to his score. If he gets a question wrong, the points for that question are subtracted from his score. Danny currently has 200 points. If he gets a 300-point question wrong, what will his score be?

a. −100

b. 0

c. −200

d. 100

54. Write 4.7×10^4 in decimal notation.

a. 4.70000

b. 47,000

c. 470,000

d. 0.00047

55. The Ravens played 25 home games this year. They had 11 losses and 2 ties. How many games did they win?

a. 12

b. 13

c. 14

d. 11

56. The temperature at midnight was 4°F. By 2 A.M. it had dropped 9°F. What was the temperature at 2 A.M.?

a. 13°F

b. −5°F

c. −4°F

d. 0°F

57. Find the next number in the following pattern.

320, 160, 80, 40, . . .

a. 35

b. 30

c. 10

d. 20

58. Which of the following terms does NOT describe the number 12?

a. prime

b. integer

c. real number

d. whole number

59. Which expression below is equal to 5?

a. $(1 + 2)^2$

b. $9 - 2^2$

c. $11 - 10 \times 5$

d. $45 \div 3 \times 3$

60. A bus picks up a group of tourists at a hotel. The sightseeing bus travels 2 blocks north, 2 blocks east, 1 block south, 2 blocks east, and 1 block south. Where is the bus in relation to the hotel?

a. 2 blocks north

b. 1 block west

c. 3 blocks south

d. 4 blocks east

61. Each week Jaime saves $25. How long will it take her to save $350?
   a. 12 weeks
   b. 14 weeks
   c. 16 weeks
   d. 18 weeks

62. Ashley's car insurance costs her $115 per month. How much does it cost her per year?
   a. $1,150
   b. $1,380
   c. $980
   d. $1,055

63. The ratio of boys to girls at the dance was 3:4. There were 60 girls at the dance. How many boys were at the dance?
   a. 45
   b. 50
   c. 55
   d. 40

Answer Explanations

1. **d.** Work backwards to find the solution. George has 5 cousins, which is 9 less than Bonnie has; therefore, Bonnie has 14 cousins. Bonnie has twice as many as Robert has, so half of 14 is 7. Robert has 7 cousins.

2. **c.** Set up a proportion with $\frac{milk}{soda}$; $\frac{2}{5} = \frac{10}{x}$. Cross multiply and solve; $(5)(10) = 2x$. Divide both sides by 2; $\frac{50}{2} = \frac{2x}{2}$; $x = 25$ sodas.

3. **b.** To earn an average of 93% on four tests, the sum of those four tests must be $(93)(4)$ or 372. The sum of the first three tests is $85 + 92 + 95 = 272$. The difference between the needed sum of four tests and the sum of the first three tests is 100. He needs a 100 to earn a 93 average.

4. **c.** To find the number of boxes needed, you should divide the number of cans by 30; $270 \div 30 = 9$ boxes.

5. **d.** From 10 A.M. Friday to 10 A.M. Saturday is 24 hours. Then, from 10 A.M. Saturday to 6 P.M. Saturday is another 8 hours. Together, that makes 32 hours.

6. **b.** Use the order of operations and try each option. The first option results in 14 because $2 \times 5 = 10$, then $10 + 4 = 14$. This does not work. The second option does result in 18. The numbers in parentheses are added first and result in 9, which is then multiplied by 2 to get a final answer of 18. Choice **c** does not work because the operation in parentheses is done first, yielding 6, which is then multiplied by 5 to get a result of 30. Choice **d** does not work because the multiplication is done first, yielding 8, which is added to 5 for a final answer of 13.

7. **a.** Multiply the number of miles (9) by the amount pledged per mile ($0.50); $9 \times 0.50 = \$4.50$. To multiply decimals, multiply normally, then count the number of decimal places in the problem and place the decimal point in the answer so that the answer has the same number of decimal places as the problem.

8. **d.** To find the square root ($\sqrt{\ }$) you ask yourself, "What number multiplied by itself gives me 64?" $8 \times 8 = 64$; therefore, 8 is the square root of 64.

9. **c.** Mr. Brown plows 6 acres an hour, so divide the number of acres (21) by 6 to find the number of hours needed; $21 \div 6 = 3.5$ hours.

10. d. This is the only answer choice that has only PRIME numbers. A prime number is a number with two and only two distinct factors. In choice **a**, 42 is not prime. In choice **b**, 4 and 6 are not prime. In choice **c**, 6 is not prime.

11. c. $2^5 = 2 \times 2 \times 2 \times 2 \times 2 = 32$

12. b. Visualize a number line. The distance from –8 to 0 is 8. Then, the distance from 0 to 63 is 63. Add the two distances together to get 71; $63 + 8 = 71$.

13. d. *Exceeded* means "gone above." Therefore, if they exceeded their goal of $9,500 by $2,100, they went over their goal by $2,100; $9,500 + $2,100 = $11,600. If you chose **a**, you subtracted $2,100 from $9,500 instead of adding the two numbers.

14. a. Subtract Mt. Kilimanjaro's height from Mt. Everest's height; $29,028 - 19,340 = 9,688$. If you chose **b**, you did not borrow correctly when subtracting.

15. a. To find the area of a square, you multiply the length of a side by itself, because all the sides are the same length. What number multiplied by itself is 36? $6 \times 6 = 36$.

16. b. Subtract the number of students present from the total number in the class to determine how many students are missing; $26 - 21 = 5$.

17. c. Divide the money raised by three to find the amount each charity will receive; $1,569 \div 3 = 523.

18. d. Find the number of points scored on two-point baskets by multiplying 2×9; 18 points were scored on two-point baskets. Find the number of points scored on three point baskets by multiplying 3×3; 9 points were scored on three-point baskets. The total number of points is the sum of these two totals; $18 + 9 = 27$.

19. a. From 2:15 P.M. to 4:15 P.M. is 2 hours. Then, from 4:15 P.M. to 4:45 P.M. is another half hour. This is a total of 2.5 hours.

20. b. There is a 1 in 6 chance of rolling a 3 because there are 6 possible outcomes on a die, but only 1 outcome is a 3.

21. a. Find the rate at which Susan is traveling by dividing her distance by time; $114 \div 2 = 57$ mph. To find out how long it will take her to travel 285 miles, divide her distance by her rate; $285 \div 57 = 5$ hours.

22. d. Divide the miles by the time to find the rate; 3,060 ÷ 5 = 612 mph.

23. c. He spent $54 on pants (3 × $18 = $54) and $120 on shirts (5 × $24 = $120). Altogether he spent $174 ($54 + $120 = $174). If you chose **a**, you calculated the cost of ONE pair of pants plus ONE shirt instead of THREE pants and FIVE shirts.

24. b. There are 100 cm in a meter. A square meter is 100 cm by 100 cm. The area of this is 10,000 sq cm (100 × 100 = 10,000).

25. a. There are 36 inches in a yard; 4 × 36 = 144 inches. There are 144 inches in 4 yards.

26. b. To find the number of hours needed to burn 1,250 calories, divide 1,250 by 500; 1,250 ÷ 500 = 2.5 hours.

27. b. Visualize a number line. The drop from 31° to 0° is 31°. There are still 9 more degrees to drop. They will be below zero. −9°F is the temperature at midnight.

28. a. Divide the total sales ($1,260) by the number of tickets sold (210) to find the cost per ticket; $1,260 ÷ 210 = $6.

29. c. A 10 second count is $\frac{1}{6}$ of a minute. To find the number of beats per minute, multiply the beat in 10 seconds by 6; 11 × 6 = 66 beats per minute.

30. c. The probability of heads does not change based on the results of previous flips. Each flip is an independent event. Therefore, the probability of getting heads is $\frac{1}{2}$.

31. b. To find the median, first put the numbers in order from least to greatest. 56, 72, 87, 89, 93. The middle number is the median. 87 is in the middle of the list, therefore, it is the median. If you chose **a**, you forgot to put the numbers in order before finding the middle number.

32. a. List the factors of 24 and 64. The largest factor that they have in common is the greatest common factor.
Factors of 24: 1, 2, 3, 4, 6, 8, 12, 24
Factors of 64: 1, 2, 4, 8, 16, 32, 64
The largest number that appears in both lists is 8.

33. d. Find the total number of slices by multiplying 3 by 8 (3 × 8 = 24). There are 24 slices to be shared among 12 coworkers. Divide the number of slices by the number of people to find the number of slices per person; 24 ÷ 12 = 2 slices per person.

34. a. Divide the number of people by the number that fit on a bus; $136 \div 51 = 2.667$. They need more than 2 buses, but not quite 3. Since you can't order part of a bus, they will need to order 3 buses.

35. d. Change all the answer choices to their decimal equivalents. Choice **a** is still 0.6; choice **b** is 0.6; choice **c** is 0.6 $(3 \div 5)$; choice **d** is 0.06; 0.06 is not equivalent to the other numbers.

36. b. Lance has 70 cents. Three-fourths of a dollar is 75 cents, so Margaret has 75 cents. Guy has 60 cents $(25 + 25 + 10 = 60)$. Bill has 60 cents $(6 \times 10 = 60)$. Margaret has the most money.

37. b. Finding what 100 students would say is the same as finding the percent, because *percent* means "out of 100." To find the percent, divide the number of students who said a dog was their favorite (258) by the total number of students surveyed (430); $258 \div 430 = 0.6$. Change 0.6 to a percent by moving the decimal two places to the right. 60%. This means that 60 out of 100 students would say dog.

38. c. Divide the bill by 5; $\$53.75 \div 5 = \$10.75$. They each pay \$10.75.

39. d. Find how much it depreciates over one year by dividing the cost by 5; $\$2,400 \div 5 = \$480$. Multiply this by 2 for two years; $\$480 \times 2 = \$960$. It will have depreciated \$960.

40. b. Add the test grades $(96 + 74 + 85 = 255)$ and divide the sum by the number of tests $(255 \div 3 = 85)$. The average is 85%.

41. b. Find the total number of people and the total number of cars. Then, divide the total people by the total cars.

People:	$57 \times 4 =$	228
	$61 \times 2 =$	122
	$9 \times 1 =$	9
	$5 \times 5 =$	25
	TOTAL	384 people

Cars: $57 + 61 + 9 + 5 = 132$

$384 \div 132 = 2.9$ which is rounded up to 3 people because 2.9 is closer to 3 than it is to 2.

42. c. Find the number of gallons per second by dividing 750 by 50 $(750 \div 50 = 15$ gallons per second). Divide 330 gallons by 15 to find how many seconds it will take $(330 \div 15 = 22$ seconds). It will take 22 seconds.

43. c. Divide 405 by 45 to get 9 minutes.

44. c. Find the probability of each event separately, and then multiply the answers. The probability of rolling a 3 is $\frac{1}{6}$ and the probability of tossing a tail is $\frac{1}{2}$. To find the probability of both of them happening, multiply $\frac{1}{6} \times \frac{1}{2} = \frac{1}{12}$. The probability is $\frac{1}{12}$.

45. c. Multiply the number of choices for each item to find the number of outfits ($5 \times 7 = 35$). There are 35 combinations.

46. c. There are 12 inches in a foot. Divide 150 by 12 to find the number of feet; $150 \div 12 = 12.5$ feet.

47. a. One cup is 8 ounces, so half a cup is 4 ounces. Multiply 15 by 4 ounces to find the number of ounces needed; $15 \times 4 = 60$ ounces.

48. c. There are 16 ounces in a pound. If Justin gains 8 ounces he will be 8 pounds and 20 ounces. The 20 ounces is 1 pound and 4 ounces. Add this to the 8 pounds to get 9 pounds and 4 ounces.

49. c. Divide the width (85 cm) by 2.54 to find the number of inches; $85 \div 2.54 = 33.46$ inches. The question says to round to the nearest tenth (one decimal place), which would be 33.5 inches.

50. b. The probability of blue is $\frac{blue}{total}$. The number of blue marbles is 6, and the total number of marbles is 16 ($3 + 6 + 5 + 2 = 16$). Therefore, the probability of choosing a blue is $\frac{6}{16} = \frac{3}{8}$.

51. c. First, put the numbers in order from least to greatest, and then find the middle of the set.

2, 2, 3, 3, 4, 4, 5, 6, 6

The middle is the average (mean) of the 5th and 6th data items. The mean of 4 and 4 is 4.

52. d. A chart like the one below can be used to determine which days Max and Ellen go to the gym. The first day after Monday that they both go—Saturday—is the answer.

today

S	M	Tu	W	Th	F	S
	M,E			E	M	
E		M	E			M,E

next day they are *both* at the gym

53. a. $200 - 300 = -100$ points

54. b. Move the decimal point 4 places to the right to get 47,000.

55. a. Thirteen games are accounted for with the losses and ties ($11 + 2 = 13$). The remainder of the 25 games were won. Subtract to find the games won; $25 - 13 = 12$ games won.

56. b. If the temperature is only 4° and drops 9°, it goes below zero. It drops 4° to zero and another 5° to –5°F.

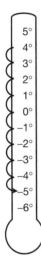

57. d. Each number is divided by 2 to find the next number; $40 \div 2 = 20$. Twenty is the next number.

58. a. Twelve is NOT prime because it has 6 factors; 1, 2, 3, 4, 6, and 12. Prime numbers have only 2 factors.

59. b. The correct order of operations must be used here. PEMDAS tells you that you should do the operations in the following order: **P**arentheses, **E**xponents, **M**ultiplication and **D**ivision—left to right, **A**ddition and **S**ubtraction—left to right.
$9 - 2^2 = 9 - 4 = 5$
a is $(1 + 2)^2 = (3)^2 = 9$
c is $11 - 10 \times 5 = 11 - 50 = -39$
d is $45 \div 3 \times 3 = 15 \times 3 = 45$

60. d. See the diagram below. They are 4 blocks east of the hotel.

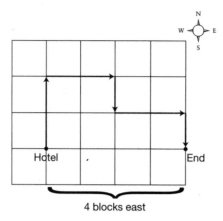

4 blocks east

61. b. Divide $350 by $25; $350 \div 25 = 14$ weeks.

62. b. Multiply $115 by 12 because there are 12 months in a year; $115 \times 12 = \$1,380$ per year.

63. a. Use a proportion comparing boys to girls at the dance.

$$\frac{boys}{girls}$$

$$\frac{3}{4} = \frac{x}{60}$$

Solve the proportion by cross-multiplying, setting the cross-products equal to each other and solving as shown below.

$(3)(60) = 4x$

$180 = 4x$

$$\frac{180}{4} = \frac{4x}{4}$$

$45 = x$

There were 45 boys.

Fractions

In order to understand arithmetic in general, it is important to practice and become comfortable with fractions and how they work. The problems in this chapter help you practice how to perform basic operations with fractions and will assist you in understanding their real-world applications.

64. Lori ran $5\frac{1}{2}$ miles Monday, $6\frac{1}{4}$ miles Tuesday, $4\frac{1}{2}$ miles Wednesday, and $2\frac{3}{4}$ miles on Thursday. What is the average number of miles Lori ran?

a. 5

b. $4\frac{1}{2}$

c. 4

d. $4\frac{3}{4}$

65. Last year Jonathan was $59\frac{5}{8}$ inches tall. This year he is $64\frac{3}{8}$ inches tall. How many inches did he grow?

a. $5\frac{1}{2}$

b. $4\frac{3}{4}$

c. $4\frac{1}{4}$

d. $5\frac{3}{4}$

66. Larry spends $\frac{3}{4}$ hour twice a day walking and playing with his dog. He also spends $\frac{1}{6}$ hour twice a day feeding his dog. How much time does Larry spend on his dog each day?

a. $\frac{11}{12}$ hour

b. $1\frac{1}{2}$ hours

c. $1\frac{5}{6}$ hours

d. $1\frac{2}{5}$ hours

67. The first section of a newspaper has 16 pages. Advertisements take up $3\frac{3}{8}$ of the pages. How many pages are not advertisements?

a. $12\frac{5}{8}$

b. $19\frac{3}{8}$

c. 13

d. $12\frac{1}{2}$

68. Lisa was assigned 48 pages to read for English class. She has finished $\frac{3}{4}$ of the assignment. How many more pages must she read?

a. 36

b. 21

c. 12

d. 8

69. Mark has three $4\frac{1}{2}$ oz cans of tomatoes and five $8\frac{1}{4}$ oz cans. How many ounces of tomatoes does Mark have?

a. $12\frac{3}{4}$

b. $54\frac{3}{4}$

c. 54

d. $62\frac{1}{4}$

70. Joe walked $2\frac{3}{4}$ miles to school, $\frac{2}{3}$ mile to work, and $1\frac{1}{4}$ miles to his friend's house. How many miles did Joe walk altogether?

a. $3\frac{6}{11}$

b. $4\frac{2}{3}$

c. $4\frac{1}{9}$

d. $3\frac{2}{3}$

71. Justin read $\frac{1}{8}$ of a book the first day, $\frac{1}{3}$ the second day, and $\frac{1}{4}$ the third day. On the fourth day he finished the book. What part of the book did Justin read on the fourth day?

   a. $\frac{2}{5}$

   b. $\frac{3}{8}$

   c. $\frac{7}{24}$

   d. $\frac{17}{24}$

72. Tammi babysat for $5\frac{1}{2}$ hours. She charged $7 an hour. How much should she get paid?

   a. $35.50

   b. $42

   c. $35

   d. $38.50

73. The painting crew has $54\frac{2}{3}$ miles of center lines to paint on the highway. If they have completed $23\frac{1}{5}$ miles, how many miles do they have to go?

   a. $31\frac{7}{15}$

   b. $31\frac{13}{15}$

   c. $21\frac{4}{15}$

   d. $31\frac{1}{2}$

74. Which of the fractions below is greater than $1\frac{1}{2}$?

   a. $\frac{7}{5}$

   b. $\frac{31}{20}$

   c. $\frac{28}{20}$

   d. $\frac{14}{11}$

75. Allison baked a cake for Guy's birthday; $\frac{4}{7}$ of the cake was eaten at the birthday party. The next day, Guy ate one third of what was left. How much of the cake did Guy eat the next day?

   a. $\frac{1}{7}$

   b. $\frac{3}{7}$

   c. $\frac{1}{4}$

   d. $\frac{4}{21}$

76. Josh practiced his clarinet for $\frac{5}{6}$ of an hour. How many minutes did he practice?

a. 83

b. 50

c. 8.3

d. 55

77. Tike Television has six minutes of advertising space every 15 minutes. How many $\frac{3}{4}$ minute commercials can they fit in the six-minute block?

a. $4\frac{1}{2}$

b. 8

c. 20

d. 11

78. Betty grew $\frac{3}{4}$ inch over the summer. Her friends also measured themselves and found that Susan grew $\frac{2}{5}$ inch, Mike grew $\frac{5}{8}$ inch, and John grew $\frac{1}{2}$ inch. List the friends in order of who grew the least to who grew the most.

a. Betty, John, Mike , Susan

b. Susan, Mike, John, Betty

c. John, Mike, Betty, Susan

d. Susan, John, Mike, Betty

79. Darma traveled 14 hours to visit her grandmother; she spent $\frac{5}{7}$ of her travel time on the highway. How many hours were not spent on the highway?

a. 3 hours

b. $2\frac{2}{7}$ hours

c. $4\frac{1}{7}$ hours

d. 4 hours

80. The Grecos are taking an $8\frac{3}{4}$ mile walk. If they walk at an average speed of $3\frac{1}{2}$ miles per hour, how long will it take them?

a. $2\frac{2}{3}$ hours

b. $30\frac{1}{8}$ hours

c. $2\frac{1}{2}$ hours

d. 5 hours

81. The town's annual budget totals $32 million. If $\frac{2}{5}$ of the budget goes toward education, how many dollars go to education?

    a. $6\frac{1}{2}$ million

    b. $9\frac{1}{5}$ million

    c. 16 million

    d. $12\frac{4}{5}$ million

82. Amy worked $\frac{4}{5}$ of the days last year. How many days did she work? (1 year = 365 days)

    a. 273

    b. 300

    c. 292

    d. 281

83. Linus wants to buy ribbon to make three bookmarks. One bookmark will be $11\frac{1}{2}$ inches long and the other two will be $8\frac{1}{4}$ inches long. How much ribbon should he buy?

    a. 27 inches

    b. 19 inches

    c. $19\frac{3}{4}$ inches

    d. 28 inches

84. Rita caught fish that weighed $3\frac{1}{4}$ lb, $8\frac{1}{2}$ lb, and $4\frac{2}{3}$ lb. What was the total weight of all the fish that Rita caught?

    a. $15\frac{4}{9}$ lb

    b. $14\frac{2}{3}$ lb

    c. $16\frac{5}{12}$ lb

    d. 15 lb

85. A rectangular garden is $4\frac{1}{2}$ yards by 3 yards. How many yards of fence are needed to surround the garden?

    a. $16\frac{1}{2}$

    b. $7\frac{1}{2}$

    c. 15

    d. 14

86. The soccer team is making pizzas for a fundraiser. They put $\frac{1}{4}$ of a pound of cheese on each pizza. If they have 12 packages of cheese, how many pizzas can they make?

    a. 48

    b. 3

    c. 24

    d. $11\frac{3}{4}$

87. Dan purchased $3\frac{1}{2}$ yards of mulch for his garden. Mulch costs $25 a yard. How much did Dan pay for his mulch?

    a. $75.00

    b. $87.50

    c. $64.25

    d. $81.60

88. Mr. Hamilton owns $2\frac{3}{4}$ acres of land. He plans to buy $1\frac{1}{3}$ more acres. How many acres will he own?

    a. $4\frac{1}{12}$

    b. $3\frac{4}{7}$

    c. $3\frac{1}{12}$

    d. 4

89. Lucy worked $32\frac{1}{2}$ hours last week and earned $195. What is her hourly wage?

    a. $7.35

    b. $5.00

    c. $6.09

    d. $6.00

90. A sheet of plywood is $5\frac{1}{2}$ feet wide and $7\frac{1}{2}$ feet long. What is the area of the sheet of plywood?

    a. $41\frac{1}{4}$ sq ft

    b. $82\frac{1}{2}$ sq ft

    c. $35\frac{1}{4}$ sq ft

    d. 11 sq ft

91. Rosa kept track of how many hours she spent reading during the month of August. The first week she read for $4\frac{1}{2}$ hours, the second week for $3\frac{3}{4}$ hours, the third week for $8\frac{1}{5}$ hours, and the fourth week for $1\frac{1}{3}$ hours. How many hours altogether did she spend reading in the month of August?

    a. $17\frac{47}{60}$

    b. 16

    c. $16\frac{1}{8}$

    d. $18\frac{2}{15}$

92. During a commercial break in the Super Bowl, there were three $\frac{1}{2}$-minute commercials and five $\frac{1}{4}$-minute commercials. How many minutes was the commercial break?

    a. $2\frac{3}{4}$

    b. $\frac{3}{4}$

    c. $3\frac{1}{2}$

    d. 5

93. Dakota's Restaurant served 635 dinners on Saturday night. On Monday night they only served $\frac{2}{5}$ as many. How many dinners did they serve on Monday?

    a. 127

    b. 254

    c. 381

    d. 385

94. Michelle is making a triple batch of chocolate chip cookies. The original recipe calls for $\frac{3}{4}$ cup of brown sugar. How many cups should she use if she is tripling the recipe?

    a. $\frac{1}{4}$

    b. 3

    c. $2\frac{1}{4}$

    d. $1\frac{3}{4}$

95. The numerator of a fraction is 4 and the denominator of the same fraction is 3. Which of the following statements is true?

    a. The value of the fraction is less than 1.

    b. The value of the fraction is greater than 1.

    c. The value of the fraction is less than 0.

    d. The value of the fraction is greater than 2.

96. The Cheshire Senior Center is hosting a bingo night; $2,400 in prize money will be given away. The first winner of the night will receive $\frac{1}{3}$ of the money. The next ten winners will receive $\frac{1}{10}$ of the remaining amount. How many dollars will each of the ten winners receive?

a. $240

b. $800

c. $160

d. $200

97. $\frac{3}{8}$ of the employees in the Acme Insurance Company work in the accounting department. What fraction of the employees does NOT work in the accounting department?

a. $\frac{10}{16}$

b. $\frac{5}{16}$

c. $\frac{3}{4}$

d. $\frac{1}{2}$

98. A bag contains 44 cookies. Kyle eats 8 of the cookies. What fraction of the cookies is left?

a. $\frac{5}{6}$

b. $\frac{9}{11}$

c. $\frac{8}{9}$

d. $\frac{3}{4}$

99. Marci's car has a 12-gallon gas tank. Her fuel gauge says that there is $\frac{1}{8}$ of a tank left. How many gallons of gas are left in the tank?

a. 3

b. $1\frac{2}{3}$

c. $\frac{3}{4}$

d. $1\frac{1}{2}$

100. Joey, Aaron, Barbara, and Stu have been collecting pennies and putting them in identical containers. Joey's container is $\frac{3}{4}$ full, Aaron's is $\frac{3}{5}$ full, Barbara's is $\frac{2}{3}$ full, and Stu's is $\frac{2}{5}$ full. Whose container has the most pennies?

a. Joey

b. Aaron

c. Barbara

d. Stu

101. What fraction of the circle below is shaded?

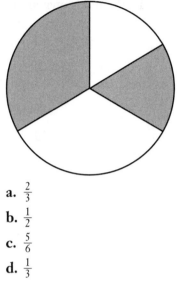

   a. $\frac{2}{3}$

   b. $\frac{1}{2}$

   c. $\frac{5}{6}$

   d. $\frac{1}{3}$

102. Michelle's brownie recipe calls for $1\frac{2}{3}$ cup of sugar. How much sugar does she need if she triples the recipe?

   a. $3\frac{2}{3}$ cups

   b. $4\frac{1}{3}$ cups

   c. $5\frac{1}{3}$ cups

   d. 5 cups

103. What fraction of the figure below is shaded?

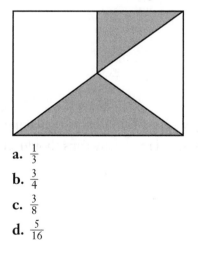

   a. $\frac{1}{3}$

   b. $\frac{3}{4}$

   c. $\frac{3}{8}$

   d. $\frac{5}{16}$

104. Mr. Reynolds owns $1\frac{3}{4}$ acres of land. He plans to buy the property next to his, which is $2\frac{1}{8}$ acres. How many acres will Mr. Reynolds own after the purchase?

　　a. $5\frac{1}{4}$

　　b. $3\frac{3}{4}$

　　c. $3\frac{1}{3}$

　　d. $3\frac{7}{8}$

105. What fraction of the figure below is shaded?

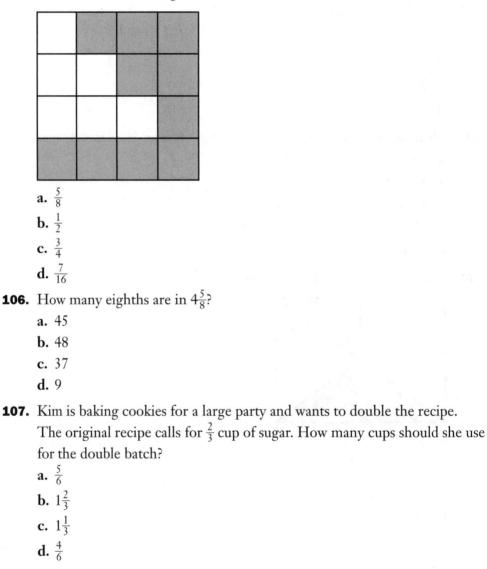

　　a. $\frac{5}{8}$

　　b. $\frac{1}{2}$

　　c. $\frac{3}{4}$

　　d. $\frac{7}{16}$

106. How many eighths are in $4\frac{5}{8}$?

　　a. 45

　　b. 48

　　c. 37

　　d. 9

107. Kim is baking cookies for a large party and wants to double the recipe. The original recipe calls for $\frac{2}{3}$ cup of sugar. How many cups should she use for the double batch?

　　a. $\frac{5}{6}$

　　b. $1\frac{2}{3}$

　　c. $1\frac{1}{3}$

　　d. $\frac{4}{6}$

108. Mr. Johnson owns $4\frac{3}{4}$ acres. He sells half of his land. How many acres does he own after the sale?

    a. $2\frac{1}{4}$

    b. $2\frac{3}{16}$

    c. $2\frac{1}{2}$

    d. $2\frac{3}{8}$

109. Which of the following is NOT equivalent to $\frac{3}{5}$?

    a. $\frac{6}{15}$

    b. 0.6

    c. $\frac{15}{25}$

    d. 60%

110. The local firefighters are doing a "fill the boot" fundraiser. Their goal is to raise $3,500. After 3 hours, they have raised $2,275. Which statement below is accurate?

    a. They have raised 35% of their goal.

    b. They have $\frac{7}{20}$ of their goal left to raise.

    c. They have raised less than $\frac{1}{2}$ of their goal.

    d. They have raised more than $\frac{3}{4}$ of their goal.

111. Lori has half a pizza left over from dinner last night. For breakfast, she eats $\frac{1}{3}$ of the leftover pizza. What fraction of the original pizza remains after Lori eats breakfast?

    a. $\frac{1}{4}$

    b. $\frac{1}{6}$

    c. $\frac{1}{3}$

    d. $\frac{3}{8}$

112. Which fraction below is closest to $\frac{1}{2}$?

    a. $\frac{2}{3}$

    b. $\frac{7}{10}$

    c. $\frac{5}{6}$

    d. $\frac{3}{5}$

113. A large coffee pot holds 120 cups. It is about three fifths full. About how many cups are in the pot?

    a. 24 cups

    b. 72 cups

    c. 48 cups

    d. 90 cups

114. An airport is backlogged with planes trying to land and has ordered planes to circle until they are told to land. A plane is using fuel at the rate of $9\frac{1}{2}$ gallons per hour, and it has $6\frac{1}{3}$ gallons left in its tank. How long can the plane continue to fly?

    a. $1\frac{1}{2}$ hours

    b. $\frac{2}{3}$ hours

    c. $2\frac{3}{4}$ hours

    d. $\frac{3}{4}$ hours

115. A carpenter receives measurements from a homeowner for a remodeling project. The homeowner lists the length of a room as $12\frac{3}{4}$ feet, but the carpenter would prefer to work in feet and inches. What is the measurement in feet and inches?

    a. 12 feet 9 inches

    b. 12 feet 8 inches

    c. 12 feet 6 inches

    d. 12 feet 3 inches

116. The state of Connecticut will pay two thirds of the cost of a new school building. If the city of New Haven is building a school that will cost a total of $15,300,000, what will the state pay?

    a. $7,750,000

    b. $5,100,000

    c. $10,200,000

    d. $1,020,000

117. One-fourth of an inch on a map represents 150 miles. The distance on the map from Springfield to Oakwood is $3\frac{1}{2}$ inches. How many miles is it from Springfield to Oakwood?

    a. 600 miles

    b. 2,100 miles

    c. 1,050 miles

    d. 5,250 miles

118. Robert brings a painting to the framing store to be framed. He chooses a frame with a 8 in by 10 in opening. The painting is $4\frac{1}{2}$ in by $6\frac{1}{2}$ in. A mat will be placed around the painting to fill the 8 in by 10 in opening. If the painting is perfectly centered, what will the width of the mat be on each side of the painting? See the diagram below.

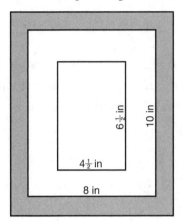

a. $3\frac{1}{2}$ in

b. $2\frac{1}{4}$ in

c. $1\frac{1}{2}$ in

d. $1\frac{3}{4}$ in

119. Mrs. Johnston's class broke into teams of three students each to play a game. The winning team received a $\frac{1}{2}$ pound jar of jellybeans. How many pounds of jellybeans will each team member get if the $\frac{1}{2}$ pound jar is shared equally among the three teammates?

a. $\frac{1}{3}$ pound

b. $\frac{1}{6}$ pound

c. $\frac{2}{5}$ pound

d. $\frac{2}{9}$ pound

120. Chuck is making a patio using $1\frac{1}{2}$ foot cement squares. The patio will be 10 cement squares by 10 cement squares. If the cement squares are placed right next to each other without any space in between, what will the dimensions of the patio be?

a. 10 ft by 10 ft

b. 20 ft by 20 ft

c. $12\frac{1}{2}$ ft by $12\frac{1}{2}$ ft

d. 15 ft by 15 ft

121. Samantha's Girl Scout troop is selling holiday wreaths. Each wreath has a bow that uses $\frac{2}{5}$ yard of ribbon. How many bows can Samantha make from a spool of ribbon that is 10 yards long?

a. 25

b. 4

c. 20

d. 8

122. Lindsay and Mark purchased a $\frac{3}{4}$ acre plot of land to build a house. Zoning laws require that houses built on less than 1 acre take up no more than half the land. What is the largest amount of land that Lindsay and Mark's house can cover?

a. $\frac{3}{8}$ acre

b. $\frac{1}{2}$ acre

c. $\frac{1}{4}$ acre

d. $\frac{5}{8}$ acre

123. Mr. Grove was watching the price of a stock he recently bought. On Monday, the stock was at $26\frac{3}{8}$. By Friday, the stock had fallen to $23\frac{1}{16}$. How much did the stock price decline?

a. $2\frac{5}{8}$

b. $3\frac{5}{16}$

c. $1\frac{3}{16}$

d. $3\frac{1}{16}$

124. Where $\frac{3}{7} = \frac{x}{42}$, what is the value of x?

a. 21

b. 6

c. 7

d. 18

125. Land in a development is selling for $60,000 per acre. If Jack purchases $1\frac{3}{4}$ acres, how much will he pay?

a. $45,000

b. $135,000

c. $105,000

d. $120,000

Answer Explanations

64. d. To find the average, add the miles run each day and divide by the number of days. To add the fractions use a common denominator of 4; $5\frac{2}{4} + 6\frac{1}{4} + 4\frac{2}{4} + 2\frac{3}{4} = 17\frac{8}{4} = 17 + 2 = 19$. Divide the sum by 4; $19 \div 4 = 4\frac{3}{4}$.

65. b. Subtract to find the difference in heights. You will need to borrow 1 whole from 64 and add it to $\frac{3}{8}$ to make the fractional part of the mixed number $\frac{11}{8}$.
$64\frac{3}{8} - 59\frac{5}{8} = 63\frac{11}{8} - 59\frac{5}{8} = 4\frac{6}{8} = 4\frac{3}{4}$

If you chose **d**, you did not borrow and simply subtracted the smaller fraction from the larger fraction.

66. c. Add the times together to find the total amount of time. Remember that he walks the dog twice and feeds the dog twice. The common denominator is 12.
$\frac{3}{4} + \frac{3}{4} + \frac{1}{6} + \frac{1}{6} = \frac{9}{12} + \frac{9}{12} + \frac{2}{12} + \frac{2}{12} = \frac{22}{12} = 1\frac{10}{12} = 1\frac{5}{6}$

If you chose **a**, you did not consider that he walks and feeds the dog TWICE a day.

67. a. Subtract the number of pages of advertisements from the total number of pages. Use a common denominator of 8 and borrow one whole ($\frac{8}{8}$) from 16 to do the subtraction.
$16 - 3\frac{3}{8} = 15\frac{8}{8} - 3\frac{3}{8} = 12\frac{5}{8}$

68. c. If Lisa has read $\frac{3}{4}$ of the assignment, she has $\frac{1}{4}$ left to go. To find $\frac{1}{4}$ of a number, divide the number by 4; $48 \div 4 = 12$ pages. If you chose **a**, you found the number of pages that she already read.

69. b. Ignore the fractional parts of the mixed numbers at first and multiply the whole number portion of the ounces by the corresponding number of cans; $4 \times 3 = 12$ ounces and $8 \times 5 = 40$ ounces. Adding together 12 and 40, you get a total of 52. Next, find the fractional portion. By multiplying the fractional part by the corresponding number of cans; $\frac{1}{2} \times 3 = \frac{3}{2} = 1\frac{1}{2}$ ounces and $\frac{1}{4} \times 5 = \frac{5}{4} = 1\frac{1}{4}$ ounces. Add together these fractional parts $1\frac{1}{2} + 1\frac{1}{4} = 2\frac{3}{4}$. Add this answer to the answer from the whole numbers to get the final answer; $2\frac{3}{4} + 52 = 54\frac{3}{4}$ ounces. If you chose **a**, you did not consider that he has THREE of the smaller cans and FIVE of the larger cans.

70. b. To find the total distance walked, add the three distances together using a common denominator of 12; $2\frac{9}{12} + \frac{8}{12} + 1\frac{3}{12} = 3\frac{20}{12}$ which is simplified to $4\frac{8}{12} = 4\frac{2}{3}$.

71. c. First, find the fraction of the book that he has read by adding the three fractions using a common denominator of 24; $\frac{3}{24} + \frac{8}{24} + \frac{6}{24} = \frac{17}{24}$. Subtract the fraction of the book he has read from one whole, using a common denominator of 24; $\frac{24}{24} - \frac{17}{24} = \frac{7}{24}$. If you chose **d**, you found the fraction of the book that Justin had already read.

72. d. Multiply the hours babysat by the charge per hour. Change the mixed number to an improper fraction before multiplying; $\frac{11}{2} \times \frac{7}{1} = \frac{77}{2}$ simplifies to $38\frac{1}{2}$ or \$38.50.

73. a. Subtract the miles completed from the total number of miles they need to paint. Use a common denominator of 15.

$$54\frac{10}{15} - 23\frac{3}{15} = 31\frac{7}{15}$$

74. b. Change the answer choices to mixed numbers to compare them to $1\frac{1}{2}$; $\frac{31}{20} = 1\frac{11}{20}$, which is larger than $1\frac{1}{2}$ because the numerator (11) is more than half the denominator (20).

75. a. Find the uneaten part of the cake by subtracting the eaten part from one whole; $\frac{3}{7}$ of the cake was uneaten. To find one third of this amount, multiply by $\frac{1}{3}$; $\frac{3}{7} \times \frac{1}{3} = \frac{1}{7}$.

76. b. An hour is 60 minutes. To find the number of minutes in $\frac{5}{6}$ of an hour, multiply 60 by $\frac{5}{6}$; $\frac{60}{1} \times \frac{5}{6} = \frac{300}{6} = 50$ minutes.

77. b. Divide the 6-minute block by $\frac{3}{4}$, remembering to take the reciprocal of the second fraction, and multiply; $6 \div \frac{3}{4} = \frac{6}{1} \times \frac{4}{3} = \frac{24}{3} = 8$.

78. d. To compare the fractions, use the common denominator of 40. Therefore, Betty = $\frac{30}{40}$, Susan = $\frac{16}{40}$, Mike = $\frac{25}{40}$, and John = $\frac{20}{40}$. To order the fractions, compare their numerators.

79. d. If $\frac{5}{7}$ of the time was spent on the highway, $\frac{2}{7}$ was not. To find $\frac{2}{7}$ of 14 hours, multiply the two numbers; $\frac{2}{7} \times \frac{14}{1} = 4$ hours.

80. c. To find the amount of time that it took the Grecos, divide the distance ($8\frac{3}{4}$) by the rate ($3\frac{1}{2}$). To divide mixed numbers, change them to improper fractions, then take the reciprocal of the second fraction and multiply; $8\frac{3}{4} \div 3\frac{1}{2} = \frac{35}{4} \div \frac{7}{2} = \frac{\overset{5}{\cancel{35}}}{\underset{2}{\cancel{4}}} \times \frac{\overset{1}{\cancel{2}}}{\underset{1}{\cancel{7}}} = \frac{5}{2} = 2\frac{1}{2}$.

81. d. To find $\frac{2}{5}$ of \$32 million, multiply the two numbers; $\frac{2}{5} \times \frac{32}{1} = \frac{64}{5}$ which simplifies to \$12$\frac{4}{5}$ million.

82. c. To find $\frac{4}{5}$ of 365 days, multiply the two numbers; $\frac{4}{5} \times \frac{365}{1} = \frac{1,460}{5}$, which simplifies to 292 days.

83. d. Add the needed lengths together using a common denominator of 4; $11\frac{2}{4} + 8\frac{1}{4} + 8\frac{1}{4} = 27\frac{4}{4}$, which simplifies to 28 inches.

84. c. Add the weights together using a common denominator of 12; $3\frac{3}{12} + 8\frac{6}{12} + 4\frac{8}{12} = 15\frac{17}{12}$, which simplifies to $16\frac{5}{12}$ lb.

85. c. Add all four sides of the garden together to find the perimeter. $4\frac{1}{2} + 4\frac{1}{2} + 3 + 3 = 14\frac{2}{2}$, which simplifies to 15 yards. If you chose **b**, you added only TWO sides of the garden.

86. a. Divide the number of packages of cheese (12) by $\frac{1}{4}$ to find the number of pizzas that can be made. Remember to take the reciprocal of the second number and multiply; $\frac{12}{1} \div \frac{1}{4} = \frac{12}{1} \times \frac{4}{1} = \frac{48}{1}$, which simplifies to 48. If you chose **b**, you multiplied by $\frac{1}{3}$ instead of dividing.

87. b. Multiply the number of yards purchased by the cost per yard. Change the mixed number into an improper fraction; $\frac{7}{2} \times \frac{25}{1} = \frac{175}{2}$, which reduces to $87\frac{1}{2}$ or \$87.50.

88. a. Add the two plots of land together using a common denominator of 12; $2\frac{9}{12} + 1\frac{4}{12} = 3\frac{13}{12}$, which simplifies to $4\frac{1}{12}$ acres.

89. d. Divide the amount of money Lucy made by the number of hours she worked. Change the mixed number to an improper fraction. When dividing fractions, take the reciprocal of the second number and multiply; $195 \div 32\frac{1}{2} = \frac{195}{1} \div \frac{65}{2} = \frac{195}{1} \times \frac{2}{65} = \frac{390}{65}$, which simplifies to \$6.

90. a. To find the area, multiply the length by the width. When multiplying mixed numbers, change the mixed numbers to improper fractions; $5\frac{1}{2} \times 7\frac{1}{2} = \frac{11}{2} \times \frac{15}{2} = \frac{165}{4}$, which simplifies to $41\frac{1}{4}$ sq ft.

91. a. Add the number of hours together using a common denominator of 60; $4\frac{30}{60} + 3\frac{45}{60} + 8\frac{12}{60} + 1\frac{20}{60} = 16\frac{107}{60}$, which simplifies to $17\frac{47}{60}$ hours.

92. a. First, multiply 3 by $\frac{1}{2}$ to find the time taken by the three half-minute commercials; $3 \times \frac{1}{2} = \frac{3}{2}$. Then, multiply $\frac{1}{4}$ by 5 to find the time taken by the five quarter-minute commercials; $\frac{1}{4} \times 5 = \frac{5}{4}$. Add the two times together to find the total commercial time. Use a common denominator of 4; $\frac{6}{4} + \frac{5}{4} = \frac{11}{4}$, which simplifies to $2\frac{3}{4}$ minutes. If you chose **b**, you only calculated for ONE commercial of each length rather than THREE $\frac{1}{2}$ minute commercials and FIVE $\frac{1}{4}$ minute commercials.

93. b. Multiply the number of Saturday dinners (635) by $\frac{2}{5}$ to find the number of dinners served on Monday; $\frac{635}{1} \times \frac{2}{5} = \frac{1,270}{5}$, which simplifies to 254 dinners.

94. c. Multiply the amount of brown sugar needed for one batch ($\frac{3}{4}$) by the number of batches (3); $\frac{3}{4} \times \frac{3}{1} = \frac{9}{4}$, which simplifies to $2\frac{1}{4}$ cups.

95. b. Since the numerator is larger than the denominator, the fraction is greater than 1.

96. c. The prize money is divided into tenths after the first third has been paid out. Find one third of $2,400 by dividing $2,400 by 3; $800 is paid to the first winner, leaving $1,600 for the next ten winners to split evenly ($2,400 − $800 = $1,600). Divide $1,600 by 10 to find how much each of the 10 winners will receive; $1,600 ÷ 10 = $160. Each winner will receive $160.

97. a. The entire company is one whole or $\frac{8}{8}$. Subtract $\frac{3}{8}$ from the whole to find the fraction of the company that is not in accounting; $\frac{8}{8} - \frac{3}{8} = \frac{5}{8}$. Recall that you subtract the numerators and leave the denominators the same when subtracting fractions; $\frac{5}{8}$ is equivalent to $\frac{10}{16}$.

98. b. If Kyle eats 8 cookies, 36 cookies are left (44 − 8 = 36). The part that is left is 26, and the whole is 44. Therefore, the fraction is $\frac{36}{44}$. Both the numerator and denominator are divisible by 4. Divide both parts by 4 to simplify the fraction to $\frac{9}{11}$.

99. d. Find $\frac{1}{8}$ of 12 gallons by multiplying; $\frac{1}{8} \times \frac{12}{1} = \frac{12}{8}$. Recall that 12 is equivalent to 12 over 1. To multiply fractions, multiply the numerators and multiply the denominators. Change $\frac{12}{8}$ to a mixed number; 8 goes into 12 once, so the whole number is 1. Four remains, so $\frac{4}{8}$ or $\frac{1}{2}$ is the fractional part; $1\frac{1}{2}$ gallons are left.

100. a. Compare $\frac{3}{4}, \frac{3}{5}, \frac{2}{3}, \frac{2}{5}$ by finding a common denominator. The common denominator for 3, 4, and 5 is 60. Multiply the numerator and denominator of a fraction by the same number so that the denominator becomes 60. The fractions then become $\frac{45}{60}, \frac{36}{60}, \frac{40}{60}, \frac{24}{60}$. The fraction with the largest numerator is the largest fraction; $\frac{45}{60}$ is the largest fraction. It is equivalent to Joey's fraction of $\frac{3}{4}$.

101. **b.** Break the circle into sixths as shown below; 3 of the sixths are shaded which is equivalent to $\frac{3}{6}$ which is $\frac{1}{2}$.

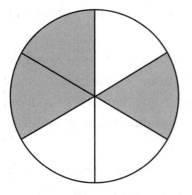

102. **d.** To triple the recipe, multiply by 3; $1\frac{2}{3} \times 3 = \frac{5}{3} \times \frac{3}{1} = \frac{15}{3} = 5$. When multiplying a mixed number, change it to an improper fraction first. To find the numerator of the improper fraction, multiply the whole number by the denominator and add the product to the numerator. Keep the denominator the same.

103. **c.** Break the block into equal regions as shown below; 3 out of the 8 blocks are shaded. This corresponds to the fraction $\frac{3}{8}$.

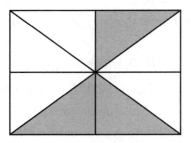

104. **d.** The two pieces of land together are $1\frac{3}{4} + 2\frac{1}{8}$ acres. Add the whole numbers $1 + 2 = 3$. Add the fractions separately writing each fraction with denominator 8: $\frac{3}{4} + \frac{1}{8} = \frac{6}{8} + \frac{1}{8} = \frac{7}{8}$. Add this to the whole number to get $3\frac{7}{8}$.

105. **a.** Ten of the 16 blocks are shaded. This is represented by the fraction $\frac{10}{16}$. Both the numerator and the denominator can be divided by 2 to simplify the fraction. This yields $\frac{5}{8}$.

106. c. Every whole has 8 eighths. Since there are 8 eighths in each of the 4 wholes, there are 32 eighths in the whole number portion. There are 5 eighths in the fraction portion. Adding the two parts together, you get 37 eighths (32 + 5 = 37).

107. c. To double the recipe, multiply by 2; $\frac{2}{3} \times 2 = \frac{2}{3} \times \frac{2}{1} = \frac{4}{3}$. Recall that 2 can be written as $\frac{2}{1}$; $\frac{4}{3}$ is an improper fraction. To change it to a mixed number, determine how many times 3 goes into 4. It goes in one time, so the whole number is 1. There is one third left over, so the mixed number is $1\frac{1}{3}$.

108. d. Divide Mr. Johnson's land by two, which is the same as multiplying by $\frac{1}{2}$; $4\frac{3}{4} \times \frac{1}{2} = \frac{19}{4} \times \frac{1}{2} = \frac{19}{8} = 2\frac{3}{8}$ acres. In the second step of the multiplication, $4\frac{3}{4}$ was changed to an improper fraction, $\frac{19}{4}$. And in the last step, the improper fraction $\frac{19}{8}$ was converted to a mixed number.

109. a. To get from $\frac{3}{5}$ to $\frac{6}{15}$ the numerator was multiplied by 2 and the denominator by 3. This does not give you an equivalent fraction. In order for the fractions to be equivalent, the numerator and denominator must be multiplied by the same number.
Choice **b** is equivalent because 3 ÷ 5 = 0.6.
Choice **c** is equivalent because the numerator and denominator of $\frac{3}{5}$ have both been multiplied by 5.
Choice **d** is equivalent because 3 ÷ 5 = 0.60, which is equivalent to 60%.

110. b. The part of their goal that they have raised is $2,275 and the whole goal is $3,500. The fraction for this is $\frac{2,275}{3,500}$. The numerator and denominator can both be divided by 175 to get a simplified fraction of $\frac{13}{20}$. They have completed $\frac{13}{20}$ of their goal, which means that they have $\frac{7}{20}$ left to go ($\frac{20}{20} - \frac{13}{20} = \frac{7}{20}$).

111. **c.** Refer to the drawing below. If half is broken into thirds, each third is one sixth of the whole. Therefore, she has $\frac{2}{6}$ or $\frac{1}{3}$ of the pizza left.

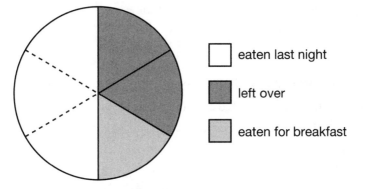

- eaten last night
- left over
- eaten for breakfast

112. **d.** The easiest way to determine which fraction is closest to $\frac{1}{2}$ is to change each of them to a decimal and compare the decimals to 0.5 which is equivalent to $\frac{1}{2}$. To find the decimal equivalents, divide the numerator by the denominator.

$\frac{2}{3} = 0.6\overline{6}$

$\frac{7}{10} = 0.7$

$\frac{5}{6} = 0.83$

$\frac{3}{5} = 0.6$

The decimal closest to 0.5 is 0.6. Therefore, $\frac{3}{5}$ is closest to $\frac{1}{2}$.

113. **b.** Multiply 120 by $\frac{3}{5}$. Thus, $\frac{120}{1} \times \frac{3}{5} = \frac{360}{5} = 72$; 72 is written as a fraction with a denominator of 1. The fraction $\frac{360}{5}$ is simplified by dividing 360 by 5 to get 72 cups.

114. **b.** Use a proportion comparing gallons used to time. The plane uses $9\frac{1}{2}$ gallons in 1 hour and the problem asks how many hours it will take to use $6\frac{1}{3}$ gallons.

$$\frac{9\frac{1}{2}\,g}{1\,hr} = \frac{6\frac{1}{3}\,g}{x\,hrs}$$

To solve the proportion for x, cross multiply, set the cross-products equal to each other and solve as shown below.

$$(9\frac{1}{2})x = (6\frac{1}{3})(1)$$

$$\frac{(9\frac{1}{2})x}{9\frac{1}{2}} = \frac{6\frac{1}{3}}{9\frac{1}{2}}$$

To divide the mixed numbers, change them into improper fractions.

$$x = \frac{\frac{19}{3}}{\frac{19}{2}}$$

$$x = \frac{19}{3} \div \frac{19}{2}$$

$$x = \frac{19}{3} \times \frac{2}{19}$$

The 19s cancel. Multiply the numerators straight across and do the same for the denominators. The final answer is $\frac{2}{3}$ hour.

115. **a.** There are 12 inches in a foot; $\frac{3}{4}$ of a foot is 9 inches ($\frac{3}{4} \times 12 = 9$). The length of the room is 12 feet 9 inches.

116. **c.** Multiply $15,300,000 by $\frac{2}{3}$; $\frac{15,300,000}{1} \times \frac{2}{3} = \frac{30,600,000}{3} = \$10,200,000$

117. **b.** Every $\frac{1}{4}$ inch on the map represents 150 miles in real life. There are 4 fourths in every whole ($4 \times 3 = 12$) and 2 fourths in $\frac{1}{2}$ for a total of 14 fourths ($12 + 2 = 14$) in $3\frac{1}{2}$ inches. Every fourth equals 150 miles. Therefore, there are 2,100 miles ($14 \times 150 = 2,100$).

118. **d.** The width of the opening of the frame is 8 inches. The painting only fills $4\frac{1}{2}$ inches of it. There is an extra $3\frac{1}{2}$ inches ($8 - 4\frac{1}{2} = 3\frac{1}{2}$) to be filled with the mat. There will be an even amount of space on either side of the painting. So, divide the extra space by 2 to find the amount of space on each side; $3\frac{1}{2} \div 2 = 1\frac{3}{4}$ inches. To divide a mixed number by a whole number, change the mixed number to an improper fraction; $3\frac{1}{2}$ becomes $\frac{7}{2}$. Also, change 2 to a fraction ($\frac{2}{1}$). Then take the reciprocal of 2 and multiply; $\frac{7}{2} \times \frac{1}{2} = \frac{7}{4} = 1\frac{3}{4}$ inches.

119. **b.** Divide $\frac{1}{2}$ pound by 3. Recall that 3 can be written as $\frac{3}{1}$; $\frac{1}{2} \div \frac{3}{1}$. When dividing fractions, take the reciprocal of the second fraction and multiply; $\frac{1}{2} \times \frac{1}{3} = \frac{1}{6}$ pound of jellybeans.

120. **d.** Multiply $1\frac{1}{2}$ by 10. Change $1\frac{1}{2}$ to an improper fraction $(\frac{3}{2})$ and make 10 into a fraction by placing it over 1 $(\frac{10}{1})$; $\frac{3}{2} \times \frac{10}{1} = \frac{30}{2} = 15$ feet. Each side is 15 feet long, so the dimensions are 15 ft by 15 ft.

121. **a.** Divide 10 by $\frac{2}{5}$. Recall that 10 can be written as $\frac{10}{1}$; $\frac{10}{1} \div \frac{2}{5}$. When dividing fractions, take the reciprocal of the second fraction and multiply; $\frac{10}{1} \times \frac{5}{2} = \frac{50}{2} = 25$ bows. If you can't figure out what operation to use in this problem, consider what you would do if you had 10 yards of ribbon and each bow took 2 yards. You would divide 10 by 2. This is the same for this problem, except that the bow takes a fraction of a yard.

122. **a.** Multiply $\frac{3}{4}$ by $\frac{1}{2}$ to find half of the land; $\frac{3}{4} \times \frac{1}{2} = \frac{3}{8}$ acre.

123. **b.** Subtract Friday's price from Monday's price; $26\frac{3}{8} - 23\frac{1}{16}$. In order to subtract, you need a common denominator. The common denominator is 16. Multiply the first fraction by 2 in the numerator and 2 in the denominator $\frac{3 \times 2}{8 \times 2} = \frac{6}{16}$. Then subtract; $26\frac{6}{16} - 23\frac{1}{16} = 3\frac{5}{16}$.

124. **d.** Determine what number 7 was multiplied by to get 42 and multiply the numerator by the same number. Seven was multiplied by six, so $3 \times 6 = 18$. The value of x is 18.

125. **c.** Multiply the cost per acre by the number of acres; $60,000 \times 1\frac{3}{4}$. Change 60,000 to a fraction by putting it over 1 and change the mixed number to an improper fraction; $\frac{60,000}{1} \times \frac{7}{4} = \frac{420,000}{4} = \$105,000$.

Decimals

We use decimals every day—to express amounts of money, or to measure distances and quantities. This chapter includes word problems that help you practice your skills in rounding decimals, as well as performing basic mathematical operations with them.

126. Round 14.853 to the nearest tenth.
    a. 14.85
    b. 10
    c. 14.9
    d. 15

127. Kara brought $23 with her when she went shopping. She spent $3.27 for lunch and $14.98 on a shirt. How much money does she have left?
    a. $8.02
    b. $4.75
    c. $18.25
    d. $7.38

128. Lucas purchased his motorcycle for $5,875.98 and sold it for $7,777.77. What was his profit?

    a. $1,901.79

    b. $2,902.89

    c. $1,051.79

    d. $1,911.89

129. Mike, Dan, Ed, and Sy played together on a baseball team. Mike's batting average was 0.349, Dan's was 0.2, Ed's was 0.35, and Sy's was 0.299. Who had the highest batting average?

    a. Mike

    b. Dan

    c. Ed

    d. Sy

130. Price Cutter sold 85 beach towels for $6.95 each. What were the total sales?

    a. $865.84

    b. $186.19

    c. $90.35

    d. $590.75

131. Rob purchased picnic food for $34.40 to share with three of his friends. They plan to split the cost evenly between the four friends. How much does each person need to pay Rob?

    a. $8.05

    b. $8.60

    c. $7.26

    d. $11.46

132. Katie ran 11.1 miles over the last three days. How many miles did she average per day?

    a. 3.7

    b. 3.0

    c. 2.4

    d. 3.3

133. Sharon purchased six adult movie tickets. She spent $52.50 on the tickets. How much was each ticket?
   a. $5.25
   b. $8.25
   c. $7.00
   d. $8.75

134. Millie purchased six bottles of soda at $1.15 each. How much did she pay?
   a. $6.15
   b. $6.90
   c. $6.60
   d. $5.90

135. Round 468.235 to the nearest hundredth.
   a. 500
   b. 568.235
   c. 468.24
   d. 468.2

136. Kenny used a micrometer to measure the thickness of a piece of construction paper. The paper measured halfway between 0.24 millimeters and 0.25 millimeters. What is the thickness of the paper?
   a. 0.05
   b. 0.245
   c. 0.255
   d. 0.3

137. In her last gymnastics competition Keri scored a 5.65 on the floor exercise, 5.85 on the vault, and 5.75 on the balance beam. What was Keri's total score?
   a. 17.25
   b. 12.31
   c. 15.41
   d. 13.5

138. Linda bought 35 yards of fencing at $4.88 a yard. How much did she spend?
   a. $298.04
   b. $248.80
   c. $91.04
   d. $170.80

139. Divide 6.8×10^5 by 2.0×10^2. Write your answer in scientific notation.
   a. 4.8×10^3
   b. $4.8 \times 10^{2.5}$
   c. 3.4×10^3
   d. 3.4×10^4

140. Leslie ordered a slice of pizza for $1.95, a salad for $2.25, and a soda for $1.05. What was the total cost of her order?
   a. $5.25
   b. $5.35
   c. $6.25
   d. $5.05

141. Last year's budget was 12.4 million dollars. This year's budget is 14.3 million dollars. How much did the budget increase?
   a. 2.5 million
   b. 1.9 million
   c. 1.4 million
   d. 2.1 million

142. Mrs. Hartill drove 3.1 miles to the grocery store, then 4.25 miles to the salon, and 10.8 miles to her son's house. How many miles did she drive altogether?
   a. 18.15
   b. 56.4
   c. 8.43
   d. 14.65

143. The following are four times from a 400-meter race. Which is the fastest time?
   a. 10.1
   b. 10.14
   c. 10.2
   d. 10.09

144. It takes the moon an average of 27.32167 days to circle the earth. Round this number to the nearest thousandth.
   a. 27.322
   b. 27.3
   c. 27.32
   d. 27.321

145. Joe's batting average is between 0.315 and 0.32. Which of the following could be Joe's average?

a. 0.311

b. 0.309

c. 0.321

d. 0.317

146. Find the product of 5.2×10^3 and 6.5×10^7. Write your answer in scientific notation.

a. 33.8×10^{21}

b. 3.38×10^{11}

c. 33.8×10^{10}

d. 3.38×10^9

147. Brian's 100-yard dash time was 2.68 seconds more than the school record. Brian's time was 13.4 seconds. What is the school record?

a. 10.72 seconds

b. 11.28 seconds

c. 10.78 seconds

d. 16.08 seconds

148. Ryan's gym membership costs him $390 per year. He pays this in twelve equal installments a year. How much is each installment?

a. $1,170

b. $42.25

c. $4,680

d. $32.50

149. How much greater is 0.0543 than 0.002?

a. 0.0343

b. 0.0072

c. 0.0523

d. 0.0563

150. Kevin ran 6.8 miles yesterday and 10.5 miles today. How many more miles did he run today?

a. 3.7

b. 4.7

c. 4.3

d. 5.9

151. Jay bought twenty-five $0.39 stamps. How much did he spend?

   a. $5.69

   b. $7.84

   c. $2.88

   d. $9.75

152. Which number falls between 5.56 and 5.81?

   a. 5.54

   b. 5.87

   c. 5.6

   d. 5.27

153. Hanna's sales goal for the week is $5,000. So far she has sold $3,574.38 worth of merchandise. How much more does she need to sell to meet her goal?

   a. $2,425.38

   b. $1,329.40

   c. $2,574.38

   d. $1,425.62

154. Which of the following decimals is the greatest number?

   a. 0.064

   b. 0.007

   c. 0.1

   d. 0.04236

155. Andy earned the following grades on his four math quizzes: 97, 78, 84, and 86. What is the average of his four quiz grades?

   a. 82.5

   b. 86.25

   c. 81.5

   d. 87

156. Luis runs at a rate of 11.7 feet per second. How far does he run in 5 seconds?

   a. 585 feet

   b. 490.65 feet

   c. 58.5 feet

   d. 55.5 feet

157. Nancy, Jennifer, Alex, and Joy ran a race. Nancy's time was 50.24 seconds, Jennifer's was 50.32, Alex's was 50.9, and Joy's was 50.2. Whose time was the fastest?
   a. Nancy
   b. Jennifer
   c. Alex
   d. Joy

158. Mike can jog 6.5 miles per hour. At this rate, how many miles will he jog in 1 hour and 30 minutes?
   a. 9.75 miles
   b. 4.7 miles
   c. 9 miles
   d. 10.75 miles

159. What decimal is represented by point A on the number line?

   a. 0.77
   b. 0.752
   c. 0.765
   d. 0.73

160. Nicole is making 20 gift baskets. She has 15 pounds of chocolates to distribute equally among the baskets. If each basket gets the same amount of chocolates, how many pounds should Nicole put in each basket?
   a. 1.3 pounds
   b. 0.8 pounds
   c. 0.75 pounds
   d. 3 pounds

161. A librarian is returning library books to the shelf. She uses the call numbers to determine where the books belong. She needs to place a book about perennials with a call number of 635.93. Between which two call numbers should she place the book?
   a. 635.8 and 635.9
   b. 635.8 and 635.95
   c. 635.935 and 635.94
   d. 635.99 and 636.0

162. Michael made 19 out of 30 free-throws this basketball season. Larry's free-throw average was 0.745 and Charles' was 0.81. John made 47 out of 86 free-throws. Who is the best free-throw shooter?

a. Michael

b. Larry

c. Charles

d. John

163. Which number below is described by the following statements? The hundredths digit is 4 and the tenths digit is twice the thousandths digit.

a. 0.643

b. 0.0844

c. 0.446

d. 0.0142

164. If a telephone pole weighs 11.5 pounds per foot, how much does a 32-foot pole weigh?

a. 368 pounds

b. 357 pounds

c. 346 pounds

d. 338.5 pounds

165. Amy purchased 6 books at $4.89 each. How much did the books cost altogether?

a. $24.84

b. $25.56

c. $29.34

d. $30.14

166. Bill traveled 117 miles in 2.25 hours. What was his average speed?

a. 26.3 miles per hour

b. 5.2 miles per hour

c. 46 miles per hour

d. 52 miles per hour

167. Tom is cutting a piece of wood to make a shelf. He cut the wood to 3.5 feet, but it is too long to fit in the bookshelf he is making. He decides to cut 0.25 feet off the board. How long will the board be after he makes the cut?

a. 3.25 feet

b. 3.75 feet

c. 3.025 feet

d. 2.75 feet

168. A bricklayer estimates that he needs 6.5 bricks per square foot. He wants to lay a patio that will be 110 square feet. How many bricks will he need?
   a. 650
   b. 7,150
   c. 6,500
   d. 715

169. Find the area of a circle with a radius of 6 inches. The formula for the area of a circle is $A = \pi r^2$. Use 3.14 for π.
   a. 37.68 square inches
   b. 113.04 square inches
   c. 9.42 square inches
   d. 75.36 square inches

170. Mary made 54 copies at the local office supply store. The copies cost $0.06 each. What was the total cost of the copies?
   a. $3.24
   b. $3.04
   c. $2.88
   d. $3.00

171. Tammi's new printer can print 13.5 pages a minute. How many pages can it print in 4 minutes?
   a. 52
   b. 48
   c. 64
   d. 54

172. The price of gasoline is $2.349 cents per gallon. If the price increases by three tenths of a cent, what will the price of gasoline be?
   a. $2.649
   b. $2.352
   c. $2.449
   d. $2.379

173. Louise is estimating the cost of the groceries in her cart. She rounds the cost of each item to the nearest dollar to make her calculations. If an item costs $1.45, to what amount will Louise round the item?
   a. $1.00
   b. $1.50
   c. $2.00
   d. $1.40

174. A pipe has a diameter of 2.5 inches. Insulation that is 0.5 inches thick is placed around the pipe. What is the diameter of the pipe with the insulation around it?

   a. 3.0 inches

   b. 4.5 inches

   c. 2.0 inches

   d. 3.5 inches

175. A 64-ounce bottle of detergent costs $3.20. What is the cost per ounce of detergent?

   a. $0.20

   b. $0.64

   c. $0.05

   d. $0.32

176. George worked from 7:00 A.M. to 3:30 P.M. with a 45-minute break. If George earns $10.50 per hour and does not get paid for his breaks, how much will he earn? (Round to the nearest cent.)

   a. $89.25

   b. $84.00

   c. $97.13

   d. $81.38

177. Marci filled her car's gas tank on Monday, and the odometer read 32,461.3 miles. On Friday when the car's odometer read 32,659.7 miles, she filled the car's tank again. It took 12.4 gallons to fill the tank. How many miles to the gallon does Marci's car get?

   a. 16 miles per gallon

   b. 18.4 miles per gallon

   c. 21.3 miles per gallon

   d. 14 miles per gallon

178. Martha has $20 to spend and would like to buy as many calculators as possible with the money. The calculators that she wants to buy are $4.25 each. How much money will she have left over after she purchases the greatest possible number of calculators?

   a. $0.75

   b. $1.25

   c. $2.75

   d. $3.00

179. What is the smallest possible number that can be created with four decimal places using the numbers 3, 5, 6, and 8?
 a. 0.8653
 b. 0.3568
 c. 0.6538
 d. 0.5368

180. Which of the following numbers is equivalent to 12.087?
 a. 12.0087
 b. 120.087
 c. 12.0870
 d. 102.087

181. Which of the following numbers will yield a number larger than 23.4 when it is multiplied by 23.4?
 a. 0.999
 b. 0.0008
 c. 0.3
 d. 1.0002

182. Kelly plans to fence in her yard. The Fabulous Fence Company charges $3.25 per foot of fencing and $15.75 an hour for labor. If Kelly needs 350 feet of fencing and the installers work a total of 6 hours installing the fence, how much will she owe the Fabulous Fence Company?
 a. $1,153.25
 b. $1,232.00
 c. $1,069.00
 d. $1,005.50

183. Thomas is keeping track of the rainfall in the month of May for his science project. The first day, 2.6 cm of rain fell. On the second day, 3.4 cm fell. On the third day, 2.1 cm fell. How many more cm are needed to reach the average monthly rainfall in May, which is 9.7 cm?
 a. 8.1 cm
 b. 0.6 cm
 c. 1.6 cm
 d. 7.4 cm

184. How will the decimal point move when 245.398 is multiplied by 1,000?
   a. It will move three places to the right.
   b. It will move three places to the left.
   c. It will move two places to the right.
   d. It will move two places to the left.

185. Mona purchased one and a half pounds of turkey at the deli for $6.90. What did she pay per pound?
   a. $4.60
   b. $10.35
   c. $3.45
   d. $5.70

186. Lucy's Lunch is sending out flyers and pays a bulk rate of 14.9 cents per piece of mail. If she mails 1,500 flyers, what will she pay?
   a. $14.90
   b. $29.80
   c. $256.50
   d. $223.50

187. If 967.234 is divided by 10, how will the decimal point move?
   a. It will move one place to the right.
   b. It will move one place to the left.
   c. It will move two places to the right.
   d. It will move two places to the left.

Answer Explanations

126. c. The tenths place is the first number to the right of the decimal. The number 8 is in the tenths place. To decide whether to round up or stay the same, look at the number to the right of the tenths place. Since that number is 5 or above, round the tenths place up to 9 and drop the digits after the tenths place.

127. b. The two items that Kara bought must be subtracted from the amount of money she had in the beginning; $23.00 − $3.27 − $14.98 = $4.75.

128. a. To find the profit, you must subtract what Lucas paid for the motorcycle from the sale price; $7,777.77 − $5,875.98 = $1,901.79.

129. c. If you add zeros to the end of Dan's and Ed's averages to make them have three decimal places, it will be easy to compare the batting averages. The four averages are: 0.349, 0.200, 0.350, and 0.299; 0.350 is the largest.

130. d. You must multiply the number of towels sold by the price of each towel; 85 × $6.95 = $590.75.

131. b. You must divide the cost of the food by 4 to split the cost evenly among the four friends; $34.40 ÷ 4 = $8.60. If you chose **d**, you divided by 3, and that does not take into account Rob's part of the bill.

132. a. To find the average number of miles, you should divide the total number of miles by the number of days; 11.1 ÷ 3 = 3.7.

133. d. To find the price of each individual ticket, you should divide the total cost by the number of tickets purchased; $52.50 ÷ 6 = $8.75.

134. b. To find the total cost of six bottles, you must multiply the cost per bottle by 6; $1.15 × 6 = $6.90.

135. c. The hundredths place is the second digit to the right of the decimal point (3). To decide how to round, you must look at the digit to the right of the hundredths place (5). Since this digit is 5 or greater, the hundredths place is rounded up to 4, producing the number 468.24.

136. b. Find the difference between 0.24 and 0.25 mm by subtracting: 0.25 − 0.24 = 0.01 mm. Half of this is 0.01 ÷ 2 = 0.005. Add to 0.24 to get 0.245 mm.

137. **a.** Keri's three scores need to be added to find the total score. To add decimals, line up the numbers and decimal points vertically and add normally; 5.65 + 5.85 + 5.75 = 17.25.

138. **d.** To multiply decimals, multiply normally, count the number of decimal places in the problem, then use the same number of decimal places in the answer; 35 × $4.88 = $170.80, since there are two decimal places in the problem, there should be two in the answer.

139. **c.** To divide numbers written in scientific notation, divide the first numbers (6.8 ÷ 2.0 = 3.4); then divide the powers of 10, which means you subtract the exponents of 10 ($10^5 ÷ 10^2 = 10^3$). The answer is $3.4 × 10^3$.

140. **a.** The cost of each item must be added together; $1.95 + $2.25 + $1.05 = $5.25.

141. **b.** Last year's budget must be subtracted from this year's budget; 14.3 million − 12.4 million = 1.9 million. Since both numbers are *millions*, the 14.3 and 12.4 can simply be subtracted and *million* is added to the answer.

142. **a.** The three distances must be added together. To add decimals, line the numbers up vertically so that the decimal points are aligned. Then, add normally; 3.1 + 4.25 + 10.8 = 18.15.

143. **d.** The fastest time is the smallest number. If you chose **c**, you chose the slowest time since it is the largest number (this person took the longest amount of time to finish the race). To compare decimals easily, make the numbers have the same number of decimal places; 10.09 < 10.10 < 10.14 < 10.20. (Note: adding zeros to the end of a number, to the right of the decimal point, does not change the value of the number.)

144. **a.** The thousandths place is the third digit to the right of the decimal point (1). To decide whether to round up or to stay the same, look at the digit to the right of the thousandths place (6). Since 6 is greater than or equal to 5, you round up to 27.322.

145. **d.** To compare decimals, you can add zeros to the end of the number after the decimal point (this will not change the value of the number); 0.315 < 0.317 < 0.320. Choice **a** is incorrect because 0.311 is smaller than 0.315. Choice **b** is incorrect because 0.309 is smaller than 0.315. Choice **c** is incorrect because 0.321 is larger than 0.32.

146. b. To multiply numbers written in scientific notation, multiply the first numbers (5.2 × 6.5 = 33.8). Then, multiply the powers of ten by adding the exponents ($10^3 \times 10^7 = 10^{10}$); 33.8×10^{10} is the answer, except it is not in scientific notation. The decimal in 33.8 must be moved to create a number between 1 and 10. Placing the decimal between the 3s will accomplish this (3.38). Since the decimal has been moved once to the left, the exponent of ten must be increased by 1. The answer is 3.38×10^{11}.

147. a. The school record is less than Brian's time. Therefore, 2.68 must be subtracted from 13.4; 13.4 − 2.68 = 10.72. To subtract decimals, line up the numbers vertically so that the decimal points are aligned. Since 13.4 has one less decimal place than 2.68, you must add a zero after the 4 (13.40) before subtracting. After you have done this, subtract normally. If you chose **d**, you added instead of subtracted.

148. d. To find each installment, the total yearly cost ($390) must be divided by the number of payments (12); 390 ÷ 12 = $32.50. Choices **a** and **c** do not make sense because they would mean that each monthly installment (payment) is more than the total yearly cost.

149. c. To find out how much greater a number is, you need to subtract; 0.0543 − 0.002 = 0.0523. To subtract decimals, line the numbers up vertically so that the decimal points align. Then, subtract normally. If you chose **a**, you did not line up the decimal places correctly. The 2 should go under the 4. If you chose **d**, you added instead of subtracted.

150. a. To find out how many more miles he ran today, subtract yesterday's miles from today's miles; 10.5 − 6.8. To subtract decimals, line the numbers up vertically so that the decimal points align. Then, subtract normally. If you chose **b**, you made an error in borrowing. You forgot to change the 10 to a 9 when borrowing 1 for the 5.

151. d. To find how much Jay spent, you must multiply the cost of each stamp ($0.39) by the number of stamps purchased (25); $0.39 × 25 = $9.75. To multiply decimals, multiply normally, then count the number of decimal places in the problem. Place the decimal point in the answer so that it contains the same number of decimal places as the problem does.

152. c. If you add a zero to the end of 5.6 to get 5.60, it is easier to see that 5.56 < 5.60 < 5.81. Choice **a** is less than 5.56. Choice **b** is greater than 5.81. Choice **d** is less than 5.56.

153. **d.** You must find the difference (subtraction) between her goal and what she has already sold. Add a decimal and two zeros to the end of $5,000 ($5,000.00) to make the subtraction easier; $5,000.00 – $3,574.38 = $1,425.62.

154. **c.** If you add zeros to the end of each of the numbers so that each number has 5 places after the decimal point, it is easier to compare the numbers; 0.00700 < 0.04236 < 0.06400 < 0.10000.

155. **b.** To find the average, you must add the items (97 + 78 + 84 + 86 = 345) and divide the sum by the total number of items (4); 345 ÷ 4 = 86.25. Remember to add a decimal point and zeros after the decimal when dividing (345.00 ÷ 4).

156. **c.** You must multiply 11.7 by 5; 11.7 × 5 = 58.5. To multiply decimals, multiply normally, then count the total number of decimal places in the problem and move the decimal point in the answer so that it contains the same number of decimal places. If you chose **a**, you forgot to add the decimal point after you multiplied. If you chose **d**, you forgot to *carry* a 3 after multiplying 7 by 5 (35, place the 5 below and carry the 3).

157. **d.** The fastest time is the smallest time. To easily compare decimals, add a zero to the end of 50.9 and 50.2 so that they read 50.90 and 50.20. Then compare the four numbers. The times are listed from smallest to largest time below.
50.20
50.24
50.32
50.90
The smallest time is 50.20 seconds.

158. **a.** One hour and thirty minutes is $1\frac{1}{2}$ hours or $\frac{3}{2}$ hours. Therefore, multiply the number of miles Mike can jog in one hour by $\frac{3}{2}$ to find the number he can jog in an hour and a half; $6.5 \times \frac{3}{2} = 9.75$ miles.

159. **a.** The hash marks indicate units of 0.01 between 0.75 and 0.80. Point A is 0.77. See the figure below.

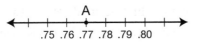

160. **c.** Nicole has 15 pounds to divide into 20 baskets. Divide 15 by 20; 15 ÷ 20 = 0.75 pounds per basket.

161. **b.** Quickly compare decimals by adding zeros to the end of a decimal so that all numbers being compared have the same number of decimal places.

Choice **a** does *not* work:

635.80

635.90

635.93—the book's call number

Choice **b** does work:

635.80

635.93—the book's call number

635.95

Choice **c** does *not* work:

635.93—the book's call number

635.935

635.94

Choice **d** does *not* work:

635.93—the book's call number

635.99

636.0

162. **c.** Change all of the comparisons to decimals by dividing the number of free-throws made by the number attempted. Michael's average is $19 \div 30 = 0.633$, John's is 0.546, Larry's was given as 0.745, and Charles' was given as 0.81. The largest decimal is the best free-throw shooter. Add zeros to the ends of the decimals to compare easily. The shooters are listed from best to worst below.

0.810 Charles

0.745 Larry

0.633 Michael

0.546 John

163. **a.** From left to right, the first decimal place is the tenths, the second is the hundredths, and the third is the thousandths. The first criterion is that the hundredths digit is 4. The second decimal place is 4, only in choice **a** and choice **c.** The second criterion is that the first decimal place is twice the third decimal place. This is only true in choice **a,** in which 6 is twice 3.

164. **a.** Multiply 11.5 by 32; $11.5 \times 32 = 368$ pounds.

165. **c.** Multiply 6 by \$4.89; $6 \times \$4.89 = \$29.34$.

166. d. Use the formula $d = rt$ (distance = rate × time). Substitute 117 miles for d. Substitute 2.25 hours for t and solve for r.

$117 = 2.25r$

$\frac{117}{2.25} = \frac{2.25r}{2.25}$

$r = 52$

The rate is 52 miles per hour.

167. a. Subtract 0.25 from 3.5; $3.5 - 0.25 = 3.25$ feet.

168. d. Multiply 6.5 by 110; $6.5 \times 110 = 715$ bricks.

169. b. Substitute 6 for r in the formula $A = \pi r^2$ and solve for A.

$A = (3.14)(6^2)$

$A = (3.14)(36)$

$A = 113.04$

The area of the circle is 113.04 square inches.

A common mistake in this problem is to say that 6^2 is 12. This is NOT true; 6^2 means 6×6 which equals 36.

170. a. Multiply 54 by $0.06 to find the total cost; $54 \times \$0.06 = \$3.24$.

171. d. Multiply 13.5 by 4 to find the number of copies made; $13.5 \times 4 = 54$ copies.

172. b. Three tenths of a cent can be written as 0.3¢, or changed to dollars by moving the decimal point two places to the left, $0.003. If $0.003 is added to $2.349, the answer is $2.352.

173. a. $1.45 is rounded to $1.00. You are rounding to the ones place, so look at the place to the right (the tenths place) to decide whether to round up or stay the same. Since 4 is less than 5, the 1 stays the same and the places after the 1 become zero.

174. **d.** The insulation surrounds the whole pipe. If the diameter is 2.5 inches, the insulation will add 0.05 inches on both sides of the diameter. See the diagram below; 2.5 + 0.5 + 0.5 = 3.5 inches.

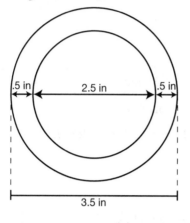

175. **c.** To find the cost per ounce, divide the cost by the number of ounces; $3.20 ÷ 64 = $0.05 per ounce.

176. **d.** First, find the number of hours George worked. From 7:00 A.M. to 3:30 P.M. is $8\frac{1}{2}$ hours. Take away his $\frac{3}{4}$ hour break and he works $7\frac{3}{4}$ hours. To find what George is paid, multiply the hours worked, 7.75 (changed from the fraction), by the pay per hour, $10.50; 7.75 × $10.50 = $81.375. The directions say to round to the nearest cent. Therefore, the answer is $81.38.

177. **a.** Since the tank was full on Monday, whatever it takes to fill the tank is the amount of gas that she has used. Therefore, she has used 12.4 gallons of gas. Next, find the number of miles Marci traveled by subtracting Monday's odometer reading from Friday's odometer reading; 32,659.7 − 32,461.3 = 198.4 miles. Divide the miles driven by the gas used to find the miles per gallon; 198.4 ÷ 12.4 = 16 miles per gallon.

178. **d.** Divide the $20 by $4.25 to find the number of calculators she can buy; $20 ÷ $4.50 = 4.405. She can buy 4 calculators. She doesn't have enough to buy a fifth calculator. This means that she has spent $17 on calculators because $4.25 × 4 = $17. To find how much she has left, subtract $20 and $17. The answer is $3.

179. **b.** Place the smallest number in the largest place value and work your way down, putting the digits in ascending order. Thus, the answer is 0.3568.

180. c. Zeros can be added to the end (right) of the decimal portion of a number without changing the value of the number; 12.0870 is equivalent to 12.087—a 0 has just been added to the end of the number.

181. d. When multiplying by a number less than 1, you get a product that is less than the number you started with. Multiplying by a number greater than 1 gives you a larger number than you started with. Therefore, multiplying by 1.0002 will yield a number larger than the one you started with.

182. b. First, figure out the cost of the fence by multiplying the number of feet of fence by $3.25; 350 × $3.25 = $1,137.50. Next, find the cost of the labor by multiplying the hours of labor by $15.75; 6 × $15.75 = $94.50. Add the two costs together to find what Kelly owes Fabulous Fence; $1,137.50 + $94.50 = $1,232.

183. c. Find the amount of rain that has fallen so far; 2.6 + 3.4 + 2.1 = 8.1 cm. Find the difference between this amount and the average rainfall by subtracting; 9.7 − 8.1 = 1.6 cm.

184. a. It is moved three places to the right. When multiplying by multiples of 10, the decimal point is moved to the right according to the number of zeros. For example: Multiply by 10 and move the decimal one place; multiply by 100 and move the decimal two places.

185. a. Divide the cost of the turkey by the weight; $6.90 ÷ 1.5 = $4.60.

186. d. Multiply the price per piece by the number of pieces; 14.9 × 1,500 = 22,350 cents. Change the cents into dollars by dividing by 100 (move the decimal point two places to the left); 22,350 cents = $223.50.

187. b. It will move one place to the left. When dividing by multiples of 10, the decimal point is moved to the left according to the number of zeros. For example: Divide by 100 and move the decimal two places; divide by 1,000 and move the decimal three places.

Percents

Percentages have many everyday uses, from figuring out the tip in a restaurant to understanding interest rates. This chapter will give you practice in solving word problems that involve percents.

188. A pair of pants costs $24. The cost was reduced by 8%. What is the new cost of the pants?
   a. $25.92
   b. $21.06
   c. $22.08
   d. $16.00

189. Michael scored 260 points during his junior year on the school basketball team. He scored 25% more points during his senior year. How many points did he score during his senior year?
   a. 195
   b. 65
   c. 325
   d. 345

190. Brian is a real estate agent. He makes a 2.5% commission on each sale. During the month of June he sold three houses. The houses sold for $153,000, $399,000, and $221,000. What was Brian's total commission on these three sales?

    a. $193,250

    b. $11,460

    c. $3,825

    d. $19,325

191. Cory purchased a frying pan that was on sale for 30% off. She saved $3.75 with the sale. What was the original price of the frying pan?

    a. $10.90

    b. $9.25

    c. $12.50

    d. $11.25

192. Peter purchased 14 new baseball cards for his collection. This increased the size of his collection by 35%. How many baseball cards does Peter now have?

    a. 5

    b. 54

    c. 40

    d. 34

193. Joey has 30 pages to read for history class tonight. He decided that he would take a break when he finished reading 70% of the pages assigned. How many pages must he read before he takes a break?

    a. 7

    b. 21

    c. 9

    d. 18

194. Julie had $500. She spent 20% of it on clothes and then 35% of the remaining money on computer equipment. How much money did Julie spend?

    a. $275

    b. $300

    c. $240

    d. $250

195. Nick paid $68.25 for a coat, including sales tax of 5%. What was the original price of the coat before tax?

    a. $63.25

    b. $65.25

    c. $65.00

    d. $64.84

196. The Dow Jones Industrial Average fell 2% today. The Dow began the day at 10,600. What was the Dow at the end of the day after the 2% drop?

    a. 10,400

    b. 10,812

    c. 10,388

    d. 7,800

197. The population of Hamden was 350,000 in 1990. By 2000, the population had decreased to 329,000. What percent of decrease is this?

    a. 16%

    b. 7.5%

    c. 6%

    d. 6.4%

198. Connecticut state sales tax is 6%. Lucy purchases a picture frame that costs $10.50. What is the Connecticut sales tax on this item?

    a. $0.60

    b. $6.30

    c. $0.63

    d. $1.05

199. Wendy brought $15 to the mall. She spent $6 on lunch. What percent of her money did she spend on lunch?

    a. 60%

    b. 40%

    c. 37%

    d. 33.3%

200. Donald sold $5,250 worth of new insurance policies last month. If he receives a commission of 7% on new policies, how much did Donald earn in commissions last month?
   a. $4,882.50
   b. $367.50
   c. $3,675.00
   d. $263.00

201. Kara borrowed $3,650 for one year at an annual interest rate of 16%. How much did Kara pay in interest?
   a. $1,168.00
   b. $584.00
   c. $4,234.00
   d. $168.00

202. Rebecca is 12.5% taller than Debbie. Debbie is 64 inches tall. How tall is Rebecca?
   a. 42 inches
   b. 8 inches
   c. 56 inches
   d. 72 inches

203. Kyra receives a 5% commission on every car she sells. She received a $1,325 commission on the last car she sold. What was the cost of the car?
   a. $26,500.00
   b. $66.25
   c. $27,825.00
   d. $16,250.00

204. A tent originally sold for $260 and has been marked down to $208. What is the percent of discount?
   a. 20%
   b. 25%
   c. 52%
   d. 18%

205. The football boosters club had 80 T-shirts made to sell at football games. By mid-October, they had only 12 left. What percent of the shirts had been sold?

a. 85%

b. 15%

c. 60%

d. 40%

206. A printer that sells for $190 is on sale for 15% off. What is the sale price of the printer?

a. $161.50

b. $175.00

c. $140.50

d. $156.50

207. What is 19% of 26?

a. 21.06

b. 4.94

c. 19

d. 5

208. There are 81 women teachers at Russell High. If 45% of the teachers in the school are women, how many teachers are there at Russell High?

a. 180

b. 36

c. 165

d. 205

209. Kim is a medical supplies salesperson. Each month she receives a 5% commission on all her sales of medical supplies up to $20,000 and 8.5% on her total sales over $20,000. Her total commission for May was $3,975. What were her sales for the month of May?

a. $79,500

b. $64,250

c. $46,764

d. $55,000

210. In the school play, 64% of the students are boys. If there are 75 students in the play, how many are boys?

   a. 64

   b. 45

   c. 27

   d. 48

211. Christie purchased a scarf marked $15.50 and gloves marked $5.50. Both items were on sale for 20% off the marked price. Christie paid 5% sales tax on her purchase. How much did she spend?

   a. $25.20

   b. $16.80

   c. $26.46

   d. $17.64

212. Last year, a math textbook cost $54. This year the cost is 107% of what it was last year. What is this year's cost?

   a. $59.78

   b. $57.78

   c. $61.00

   d. $50.22

213. Larry earned $32,000 per year. Then he received a $3\frac{1}{4}$% raise. What is Larry's salary after the raise?

   a. $33,040

   b. $35,000

   c. $32,140

   d. $32,960

214. Bill spent 50% of his savings on school supplies, then he spent 50% of what was left on lunch. If he had $8 left after lunch, how much did he have in savings at the beginning?

   a. $32

   b. $16

   c. $30

   d. $18

215. A coat that costs $72 is marked up 22%. What is the new price of the coat?
   a. $78.22
   b. $96.14
   c. $56.16
   d. $87.84

216. Kristen earns $550 each week after taxes. She deposits 10% of her income in a savings account and 7% in a retirement fund. How much does Kristen have left after the money is taken out for her savings account and retirement fund?
   a. $505.25
   b. $435.50
   c. $533.00
   d. $456.50

217. Coastal Cable had 1,440,000 customers in January of 2002. During the first half of 2002 the company launched a huge advertising campaign. By the end of 2002 they had 1,800,000 customers. What is the percent of increase?
   a. 36%
   b. 21%
   c. 20%
   d. 25%

218. The price of heating oil rose from $1.70 per gallon to $2.38 per gallon. What is the percent of increase?
   a. 40%
   b. 33%
   c. 29%
   d. 38%

219. 450 girls were surveyed about their favorite sport: 24% said that basketball is their favorite sport, 13% said that ice hockey is their favorite sport, and 41% said that softball is their favorite sport. The remaining girls said that field hockey is their favorite sport. What percent of the girls surveyed said that field hockey is their favorite sport?
   a. 37%
   b. 22%
   c. 78%
   d. 35%

220. At Yale New Haven Hospital, 25% of babies born weigh less than 6 pounds and 78% weigh less than 8.5 pounds. What percent of the babies born at Yale New Haven Hospital weigh between 6 and 8.5 pounds?
   a. 22%
   b. 24%
   c. 53%
   d. 2.5%

221. An $80.00 coat is marked down 20%. It does not sell, so the shop owner marks it down an additional 15%. What is the new price of the coat?
   a. $64.00
   b. $68.60
   c. $52.00
   d. $54.40

222. $\frac{3}{5}$ of the soda purchased at the football game was cola. What percentage of the soda purchased was cola?
   a. 60%
   b. 40%
   c. 30%
   d. 70%

223. In a recent survey of 700 people, 18% said that red was their favorite color. How many people said that red was their favorite color?
   a. 18
   b. 88
   c. 125
   d. 126

224. In Kimmi's fourth grade class, 8 out of the 20 students walk to school. What percent of the students in her class walk to school?
   a. 40%
   b. 50%
   c. 45%
   d. 35%

225. In Daniel's fifth grade class, 37.5% of the 24 students walk to school. One-third of the walkers got a ride to school today from their parents. How many walkers got a ride to school from their parents today?

    a. 9

    b. 12

    c. 2

    d. 3

226. Lindsay purchased a pocketbook for $45 and a pair of shoes for $55. The sales tax on the items was 6%. How much sales tax did she pay?

    a. $2.70

    b. $3.30

    c. $6.00

    d. $6.60

227. Wendy bought a book and the sales tax on the book was $2.12. If the sales tax is 8%, what was the price of the book?

    a. $26.50

    b. $16.96

    c. $24.76

    d. $265.00

228. Mr. Pelicas took his family out to dinner. The bill was $65.00. He would like to leave a 17% tip. How much should he leave?

    a. $17

    b. $3.25

    c. $11.05

    d. $16.25

229. The *Daily News* reported that 54% of people surveyed said that they would vote for Larry Salva for mayor. Based on the survey results, if 23,500 people vote in the election, how many people are expected to vote for Mr. Salva?

    a. 12,690

    b. 4,350

    c. 10,810

    d. 18,100

230. Bikes are on sale for 30% off the original price. What percent of the original price will the customer pay if he gets the bike at the sale price?

    a. 130%

    b. 60%

    c. 70%

    d. 97%

231. A pair of mittens has been discounted 13.5%. The original price of the mittens was $10. What is the new price?

    a. $8.80

    b. $8.65

    c. $7.50

    d. $9.87

232. John's youth group is trying to raise $1,500 at a car wash. So far, they have raised $525. What percent of their goal have they raised?

    a. 35%

    b. 52%

    c. 3%

    d. 28%

233. The freshman class is participating in a fundraiser. Their goal is to raise $5,000. After the first two days of the fundraiser, they have raised 32% of their goal. How many dollars did they raise the first two days?

    a. $160

    b. $32

    c. $1,600

    d. $3,400

234. 1,152 out of 3,600 people surveyed said that they work more than 40 hours per week. What percent of the people surveyed said that they work more than 40 hours per week?

    a. 32%

    b. 3.1%

    c. 40%

    d. 24%

235. Peter was 60 inches tall on his thirteenth birthday. By the time he turned 15, his height had increased 15%. How tall was Peter when he turned 15?
a. 75 inches
b. 69 inches
c. 72 inches
d. 71 inches

236. Laura paid $17 for a pair of jeans. The ticketed price was 20% off the original price plus the sign on the rack said, "Take an additional 15% off the ticketed price." What was the original price of the jeans?
a. $30
b. $28
c. $25
d. $21

237. The 6% sales tax on a basket was $0.72. What was the price of the basket?
a. $12
b. $43.20
c. $17
d. $24.50

238. What percent of the figure below is shaded?

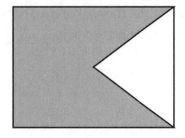

a. 50%
b. 65%
c. 75%
d. 80%

239. Melisa and Jennifer threw a fiftieth birthday party for their father at a local restaurant. When the bill came, Melisa added a 15% tip of $42. Jennifer said that the service was wonderful and they should leave a 20% tip instead. How much is a 20% tip?
a. $56
b. $45
c. $47
d. $60

240. The Hamden Town Manager wants to know what percent of the snow removal budget has already been spent. The budget for snow removal is $130,000. It has been an exceptionally snowy year, and they have already spent $100,000 for snow removal. What percent of the budget has already been spent? (round to the nearest percent)
a. 30%
b. 70%
c. 72%
d. 77%

241. A real estate agent makes a 1.5% commission on her sales. What is her commission if she sells a $359,000 house?
a. $53,850
b. $5,385
c. $23,933
d. $1,500

242. The manager of a specialty store marks up imported products 110%. If a vase imported from Italy costs him $42, what price tag will he put on the item?
a. $84
b. $83.50
c. $65
d. $88.20

243. Michelle purchased a vacation home with her sisters. Michelle has $125,000 invested in the property, which is worth $400,000. What percent of the property does Michelle own?
a. 3.2%
b. 43%
c. 31.25%
d. 26.5%

244. Kyra's weekly wages are $895. A Social Security tax of 7.51% and a State Disability Insurance of 1.2% are taken out of her wages. What is her weekly paycheck, assuming there are no other deductions?
a. $827.79
b. $884.26
c. $962.21
d. $817.05

245. Oscar's Oil Company gives customers a 5% discount if they pay their bill within 10 days. The Stevens' oil bill is $178. How much do they save if they pay the bill within 10 days?
a. $8.90
b. $5.00
c. $17.80
d. $14.60

246. Josephine is on an 1,800 calorie per day diet. She tries to keep her intake of fat to no more than 30% of her total calories. Based on an 1,800 calorie a day diet, what is the maximum number of calories that Josephine should consume from fats per day to stay within her goal?
a. 600
b. 640
c. 580
d. 540

247. A family may deduct 24% of their childcare expenses from their income tax owed. If a family had $1,345 in childcare expenses, how much can they deduct?
a. $1,022.20
b. $345.00
c. $322.80
d. $789.70

248. A factory that is working at 90% capacity is shipping 450 cars per week. If the factory works at 100% capacity, how many cars can it ship per week?

   a. 650
   b. 500
   c. 495
   d. 405

249. Sales increased by only $\frac{1}{2}$% last month. If the sales from the previous month were $152,850, what were last month's sales?

   a. $229,275.00
   b. $153,614.25
   c. $152,849.05
   d. $151,397.92

250. Laura is planning her wedding. She expects 220 people to attend the wedding, but she has been told that approximately 5% typically don't show. About how many people should she expect not to show?

   a. 46
   b. 5
   c. 11
   d. 23

Answer Explanations

188. c. If the cost of the pants is reduced by 8%, the cost of the pants is 92% of the original cost (100% − 8% = 92%). To find 92% of the original cost, multiply the original cost of the pants by the decimal equivalent of 92%; $24 × 0.92 = $22.08.

189. c. If the number of points is increased by 25%, the number of points in his senior year is 125% of the number of points in his junior year (100% + 25% = 125%). To find 125% of the number of points in his junior year, multiply the junior year points by the decimal equivalent of 125%; 260 × 1.25 = 325. If you chose **a**, you calculated what his points would be if he scored 25% LESS than he did in his junior year.

190. d. First, find the total of Brian's sales; $153,000 + $399,000 + $221,000 = $773,000. To find 2.5% of $773,000, multiply by the decimal equivalent of 2.5%; $773,000 × 0.025 = $19,325. If you chose **a**, you used the decimal 0.25, which is 25%, NOT 2.5%.

191. c. Use a proportion to find the original cost of the frying pan; $\frac{part}{whole} = \frac{\%}{100}$. The $3.75 that was saved is *part* of the original price. The *whole* price is what we are looking for, so call it x. The % is 30 (the percent off); $\frac{3.75}{x} = \frac{30}{100}$. To solve the proportion, cross-multiply. (3.75)(100) = 30x. Divide both sides by 30 to solve for x; $\frac{375}{30} = \frac{30x}{30}$; $x = $12.50.

192. b. First, you must find how many baseball cards Peter had originally. Use a proportion to find the original number of baseball cards; $\frac{part}{whole} = \frac{\%}{100}$. The 14 baseball cards that he added to his collection is the *part*. The *whole* number of baseball cards is what we are looking for, so call it x. The % is 35 (the percent of increase); $\frac{14}{x} = \frac{35}{100}$. To solve the proportion, cross-multiply; (14)(100) = 35x. Divide both sides by 35 to solve for x; $\frac{1,400}{35} = \frac{35x}{35}$; $x = 40$. The original number of baseball cards was 40, and 14 more were added to the collection for a total of 54 cards.

193. b. To find 70% of 30, you must multiply 30 by the decimal equivalent of 70% (0.70); 30 × 0.70 = 21. If you chose **c**, you calculated how many pages he has left to read after his break.

194. c. Find 20% of $500 by multiplying $500 by the decimal equivalent of 20% (0.20); $500 × 0.20 = $100. She spent $100 on clothes, leaving her with $400. Find 35% of $400; 0.35 × 400 = $140. Julie spent $140 on computer equipment; $100 on clothes plus $140 on computer equipment totals $240 spent. If you chose **a**, you found 55% (20% + 35%) of the total without taking into account that the 35% was on the amount of money Julie had AFTER spending the original 20%.

195. c. Since 5% sales tax was added to the cost of the coat, $68.25 is 105% of the original price of the coat. Use a proportion to find the original cost of the coat; $\frac{part}{whole} = \frac{\%}{100}$. *Part* is the price of the coat with the sales tax, $68.25. *Whole* is the original price on the coat that we are looking for. Call it x. The % is 105; $\frac{68.25}{x} = \frac{105}{100}$. To solve for x, cross-multiply; (68.25)(100) = 105x. Divide both sides by 105; $\frac{6,825}{105} = \frac{105x}{105}$; x = $65.00.

196. c. The Dow lost 2%, so it is worth 98% of what it was worth at the beginning of the day (100% − 2% = 98%). To find 98% of 10,600, multiply 10,600 by the decimal equivalent of 98%; 10,600 × 0.98 = 10,388.

197. c. First, find the number of residents who left Hamden by subtracting the new population from the old population; 350,000 − 329,000 = 21,000. The population decreased by 21,000. To find what percent this is of the original population, divide 21,000 by the original population of 350,000; 21,000 ÷ 350,000 = 0.06; 0.06 is equivalent to 6%. If you chose **d**, you found the decrease in relation to the NEW population (2000) when the decrease must be in relation to the original population (1990).

198. c. Find 6% of $10.50 by multiplying $10.50 by 0.06 (the decimal equivalent of 6%); $10.50 × 0.06 = $0.63. If you chose **b**, you found 60% (0.6) instead of 6% (0.06).

199. b. Divide $6 by $15 to find the percent; $6 ÷ $15 = 0.40; 0.40 is equivalent to 40%.

200. b. To find 7% of $5,250, multiply $5,250 by the decimal equivalent of 7% (0.07); $5,250 × 0.07 = $367.50.

201. b. To find 16% of $3,650, multiply $3,650 by the decimal equivalent of 16% (0.16); $3,650 × 0.16 = $584.

202. d. Since Rebecca is 12.5% taller than Debbie, she is 112.5% of Debbie's height (100% + 12.5% = 112.5%). To find 112.5% of Debbie's height, multiply Debbie's height by the decimal equivalent of 112.5% (1.125); 64 × 1.125 = 72 inches. If you chose **c**, you found what Rebecca's height would be if she were 12.5% SHORTER than Debbie (you subtracted instead of added).

203. a. Use the proportion $\frac{part}{whole} = \frac{\%}{100}$ to solve the problem; $1,325 is the *part* and 5% is the %. We are looking for the *whole* so we will call it x; $\frac{1,325}{x} = \frac{5}{100}$. Cross multiply; (1,325)(100) = 5x. Divide both sides by 5 to solve for x; $\frac{132,500}{5} = \frac{5x}{5}$; x = $26,500. If you chose **b**, you found 5% of her commission (5% of $1,325).

204. a. Find the number of dollars off. $260 − $208 = $52. Next, determine what percent of the original price $52 is by dividing $52 by the original price, $260; $52 ÷ $260 = 0.20; 0.20 is equivalent to 20%.

205. a. Determine the number of T-shirts sold; 80 − 12 = 68. To find what percent of the original number of shirts 68 is, divide 68 by 80; 68 ÷ 80 = 0.85; 0.85 is equivalent to 85%. If you chose **b**, you found the percent of T-shirts that were LEFT instead of the percent that had been SOLD.

206. a. The printer is 15% off. That means that it is 85% of its original price (100% − 15% = 85%). To find 85% of $190, multiply $190 by the decimal equivalent of 85% (0.85); $190 × 0.85 = $161.50.

207. b. To find 19% of 26, multiply 26 by the decimal equivalent of 19% (0.19); 26 × 0.19 = 4.94.

208. a. Use the proportion $\frac{part}{whole} = \frac{\%}{100}$. *Part* is the number of female teachers (81). *Whole* is what we are looking for; call it x; the % is 45; $\frac{81}{x} = \frac{45}{100}$. Cross multiply; (81)(100) = 45x. Divide both sides by 45 to solve for x; $\frac{8,100}{45} = \frac{45x}{45}$; x = 180.

209. d. Kim sold over $20,000 in May. She received a 5% commission on the first $20,000 of sales. To find 5%, multiply by the decimal equivalent of 5% (0.05); $20,000 × 0.05 = $1,000. Since her total commission was $3,975, $3,975 − $1,000 = $2,975 is the amount of commission she earned on her sales over $20,000. $2,975 is 8.5% of her sales over $20,000. To find the amount of her sales over $20,000, use a proportion; $\frac{part}{whole} = \frac{\%}{100}$. *Part* is $2,975, and *whole* is what we are looking for, so let's call it x. The % is 8.5; $\frac{2,975}{x} = \frac{8.5}{100}$. To solve for x, cross multiply; (2,975)(100) = 8.5x. Divide both sides by 8.5 to solve; $\frac{297,500}{8.5} = \frac{8.5x}{8.5}$; x = $35,000. Her sales over $20,000 were $35,000. Her total sales were $55,000 ($20,000 + $35,000).

210. **d.** To find 64% of 75, multiply 75 by the decimal equivalent of 64% (0.64); 75 × 0.64 = 48. If you chose **c**, you found the number of girls.

211. **d.** First, find the sale price of the scarf and the gloves. They are both 20% off, which means that Christie paid 80% of the original price (100% − 20% = 80%). To find 80% of each price, multiply the price by the decimal equivalent of 80% (0.80); $15.50 × 0.80 = $12.40. $5.50 × 0.80 = $4.40. Together the two items cost $16.80 ($12.40 + $4.40 = $16.80). There is 5% sales tax on the total price. To find 5% of $16.80, multiply $16.80 by the decimal equivalent of 5% (0.05); $16.80 × 0.05 = $0.84. The tax is $0.84. Christie paid a total of $17.64 ($16.80 + $0.84 = $17.64).

212. **b.** To find 107% of $54, multiply $54 by the decimal equivalent of 107% (1.07); $54 × 1.07 = $57.78. If you chose **d**, you found what the cost of the book would be if it was 7% LESS next year.

213. **a.** If Larry earns a $3\frac{1}{4}$% (or 3.25%) raise, he will earn 103.25% of his original salary. To find 103.35% of $32,000, multiply $32,000 by the decimal equivalent of 103.25% (1.0325); $32,000 × 1.0325 = $33,040. If you chose **d**, you found his salary with a 3% raise when multiplying by 1.03 or 0.03 and then adding that answer to his original salary.

214. **a.** Work backwards to find the answer. After lunch Bill had $8. He had spent 50% (or $\frac{1}{2}$) of what he had on lunch and 50% is what is left. Since $8 is 50% of what he had before lunch, he had $16 before lunch. Using the same reasoning, $16 is 50% of what he had before buying school supplies. Therefore, he had $32 when he began shopping.

215. **d.** If a coat is marked up 22%, it is 122% of its original cost (100% + 22% = 122%). To find 122% of the original cost, multiply $72 by the decimal equivalent of 122% (1.22); $72 × 1.22 = $87.84.

216. **d.** Kristen has a total of 17% taken out of her check. Therefore, she is left with 83% of what she started with (100% − 17% = 83%). To find 83% of $550, multiply $550 by the decimal equivalent of 83%; $550 × 0.83 = $456.50.

217. d. Coastal Cable gained a total of 360,000 customers (1,800,000 − 1,440,000 = 360,000). To find out what percent of the original number of customers 360,000 represents, divide 360,000 by 1,440,000; 360,000 ÷ 1,440,000 = 0.25; 0.25 is equivalent to 25%. If you chose **c**, you found the percent of increase in relation to the new number of customers (1,800,000) rather than the original number of customers (1,440,000).

218. a. The price of heating oil rose $0.68 ($2.38 − $1.70 = $0.68). To find the percent of increase, divide $0.68 by the original cost of $1.70; $0.68 ÷ $1.70 = 0.4; 0.4 is equivalent to 40%. If you chose **c**, you found the percent of increase in relation to the new price ($2.38) rather than the original price ($1.70).

219. b. The percents must add to 100%; 24% + 13% + 41% = 78%. If 78% of the girls surveyed have been accounted for, the remainder of the girls must have said that field hockey is their favorite sport. To find the percent that said field hockey is their favorite sport, subtract 78% from 100%; 100% − 78% = 22%; 22% of the girls said that field hockey is their favorite sport.

220. c. 78% weigh less than 8.5 pounds, but you must subtract the 25% that are below 6 pounds; 78% − 25% = 53%. 53% of the babies weigh between 6 and 8.5 pounds.

221. d. Find 20% of the original price of the coat and subtract it from the original price. To find 20%, multiply by 0.20; $80 × 0.20 = $16. Take $16 off the original price; $80 − $16 = $64. The first sale price is $64. Take 15% off this using the same method; $64 × 0.15 = $9.60; $64 − $9.60 = $54.40. The new price of the coat is $54.40.

   Another way of solving this problem is to look at the percent that is left after the discount has been taken. For example, if 20% is taken off, 80% is left (100% − 20%). Therefore, 80% of the original price is $80 × 0.80 = $64. If 15% is taken off this price, 85% is left; $64 × 0.85 = $54.40. This method eliminates the extra step of subtracting.

222. a. Change the fraction to a decimal by dividing the numerator by the denominator (top ÷ bottom); 3 ÷ 5 = 0.6. Change 0.6 to a percent by multiplying by 100; 0.6 × 100 = 60%. Recall that multiplying by 100 means that the decimal point is moved two places to the right.

223. d. Find 18% of 700 by multiplying 700 by the decimal equivalent of 18% (0.18); $700 \times 0.18 = 126$; 126 people said that red is their favorite color.

Another way of looking at this problem is to recall that 18% means "18 out of 100." Since 700 is 7 times 100, multiply 18 by 7 to find the number of people out of 700 who said red was their favorite color; $18 \times 7 = 126$.

224. a. Write the relationship as a fraction; $\frac{part}{whole}$ or $\frac{walkers}{total} = \frac{8}{20}$. Find the decimal equivalent by dividing the numerator by the denominator (top ÷ bottom); $8 \div 20 = 0.4$. Change 0.4 to a percent by multiplying by 100; $0.4 \times 100 = 40\%$. Recall that multiplying by 100 means that the decimal point is moved two places to the right.

Another way to look at this problem is using a proportion; $\frac{part}{whole} = \frac{\%}{100}$. You are looking for the percent, so that will be the variable; $\frac{8}{20} = \frac{x}{100}$. To solve the proportion, cross-multiply and set the answers equal to each other; $(8)(100) = 20x$. Solve for x by dividing both sides by 20.
$$800 = 20x$$
$$\frac{800}{20} = \frac{20x}{20}$$
$$x = 40$$
40% of the students are walkers.

225. d. First, find the number of walkers and then find one third of that number. Find 37.5% of 24 by multiplying 24 by the decimal equivalent of 37.5%. To find the decimal equivalent, move the decimal point two places to the left; $37.5\% = 0.375$. Now, multiply $24 \times 0.375 = 9$. Find one third of 9 by dividing 9 by 3; $9 \div 3 = 3$. Three walkers got rides.

226. c. Find the price of the two items together (without tax); $45 + $55 = $100. Next, find 6% of $100. You can multiply $100 by 0.06, but it is easier to realize that 6% means "6 out of 100," so 6% of $100 is $6. The sales tax is $6.

A common mistake is to use 0.6 for 6% instead of 0.06; 0.6 is 60%. To find the decimal equivalent of a percent, you must move the decimal point two places to the left.

227. a. A proportion can be used to solve this problem; $\frac{part}{whole} = \frac{\%}{100}$. In this example, the *part* is the tax, the % is 8, and the *whole* is x. To solve the proportion, cross-multiply, set the cross-products equal to each other, and solve as shown below.

$$\frac{2.12}{x} = \frac{8}{100}$$

$(2.12)(100) = 8x$

$212 = 8x$

$$\frac{212}{8} = \frac{8x}{8}$$

$x = 26.5$

The price of the book is $26.50.

228. c. Find 17% by multiplying $65 by the decimal equivalent of 17% (0.17); $65 × 0.17 = $11.05. The tip is $11.05.

229. a. Find 54% of 23,500 by multiplying 23,500 by the decimal equivalent of 54% (0.54); 23,500 × 0.54 = 12,690; 12,690 people are expected to vote for Mr. Salva.

230. c. The original price of the bike is 100%. If the sale takes 30% off the price, it will leave 70% of the original price (100% − 30% = 70%).

231. b. Find 13.5% of $10 and subtract it from $10. Find 13.5% of $10 by multiplying $10 by the decimal equivalent of 13.5% (0.135); $10 × 0.135 = $1.35; $1.35 is taken off the price of the mittens. Subtract $1.35 from $10 to find the sale price; $10 − $1.35 = $8.65. The sale price is $8.65.

Another way to compute the sale price is to find what percent is left after taking the discount. The original price was 100% and 13.5% is taken off; 86.5% is left (100% − 13.5% = 86.5%). Find 86.5% of the original cost by multiplying $10 by the decimal equivalent of 86.5% (0.865); $10 × 0.865 = $8.65.

232. **a.** Use a proportion to solve the problem; $\frac{part}{whole} = \frac{\%}{100}$. The *whole* is $1,500 and the *part* is $525. You are looking for the %, so it is x. To solve the proportion, cross-multiply, set the cross-products equal to each other, and solve as shown below.

$$\frac{525}{1,500} = \frac{x}{100}$$
$$(1,500)x = (525)(100)$$
$$1,500x = 52,500$$
$$\frac{1,500x}{1,500} = \frac{52,500}{1,500}$$
$$x = 35\%$$

They have raised 35% of the goal.

Another way to find the percent is to divide the part by the whole, which gives you a decimal. Convert the decimal into a percent by multiplying by 100 (move the decimal point two places to the right); $\frac{525}{1,500} = 0.35 = 35\%$.

233. **c.** Find 32% of $5,000 by multiplying $5,000 by the decimal equivalent of 32% (0.32); $5,000 × 0.32 = $1,600.

234. **a.** Divide the part by the whole; $1,152 ÷ 3,600 = 0.32$. Change the decimal to a percent by multiplying by 100 (move the decimal point two places to the right); 32% of the people surveyed said that they work more than 40 hours a week.

Another way to find the answer is to use a proportion; $\frac{part}{whole} = \frac{\%}{100}$. The *part* is 1,152, the *whole* is 3,600, and the % is x. To solve the proportion, cross-multiply, set the cross-products equal to each other, and solve as shown below.

$$\frac{1,152}{3,600} = \frac{x}{100}$$
$$3,600x = (1,152)(100)$$
$$3,600x = 115,200$$
$$\frac{3,600x}{3,600} = \frac{115,200}{3,600}$$
$$x = 32$$

235. **b.** Find 15% of 60 inches and add it to 60 inches. Find 15% by multiplying 60 by the decimal equivalent of 15% (0.15); 60 × 0.15 = 9. Add 9 inches to 60 inches to get 69 inches.

236. c. Call the original price of the jeans x. First 20% is deducted from the original cost (the original cost is 100%); 80% of the original cost is left (100% – 20% = 80%); 80% of x is 0.80x. The cost of the jeans after the first discount is 0.80x. This price is then discounted 15%. Remember 15% is taken off the discounted price; 85% of the discounted price is left. Multiply the discounted price by 0.85 to find the price of the jeans after the second discount; (0.85)(0.80x) is the cost of the jeans after both discounts. We are told that this price is $17. Set the two expressions for the cost of the jeans equal to each other (0.85)(0.80x) = $17 and solve for x (the original cost of the jeans).

(0.85)(0.80x) = 17

0.68x = 17

$\frac{0.68x}{0.68} = \frac{17}{0.68}$

$x = 25$

The original price of the jeans was $25.

237. a. Use a proportion to solve the problem; $\frac{part}{whole} = \frac{\%}{100}$. The *whole* is the price of the basket (which is unknown, so call it x), the *part* is the tax of $0.72, and the *percentage* is 5. The proportion is $\frac{0.72}{x} = \frac{6}{100}$. Solve the proportion by cross-multiplying, setting the cross-products equal to each other, and solving as shown below.

$\frac{0.72}{x} = \frac{6}{100}$

(100)(0.72) = 6x

72 = 6x

$\frac{72}{6} = \frac{6x}{6}$

12 = x

The price of the basket was $12.

238. c. Break the rectangle into eighths as shown below. The shaded part is $\frac{6}{8}$ or $\frac{3}{4}$; $\frac{3}{4}$ is 75%.

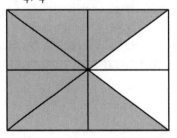

239. a. To find 20%, add 5% to 15%. Since 15% is known to be $42, 5% can be found by dividing $42 by 3 (15% ÷ 3 = 5%); $42 ÷ 3 = $14. To find 20%, add the 5% ($14) to the 15% ($42); $14 + $42 = $56; 20% is $56.

240. d. Use a proportion to solve the problem; $\frac{part}{whole} = \frac{\%}{100}$. The *part* is $100,000, the *whole* is $130,000, and the *percentage* is x because it is unknown; $\frac{100,000}{130,000} = \frac{x}{100}$. To solve the proportion, cross-multiply, set the cross-products equal to each other, and solve as shown below.

$$\frac{100,000}{130,000} = \frac{x}{100}$$

$$(100)(100,000) = 130,000x$$

$$10,000,000 = 130,000x$$

$$\frac{10,000,000}{130,000} = \frac{130,000x}{130,000}$$

$$x = 77$$

77% of the budget has been spent.

241. b. Multiply $359,000 by the decimal equivalent of 1.5% (0.015) to find her commission; $359,000 × 0.015 = $5,385; $5,385 is the commission.

A common mistake is to use 0.15 for the decimal equivalent of 1.5%; 0.15 is equivalent to 15%. Remember, to find the decimal equivalent of a percent, move the decimal point two places to the left.

242. d. To find the price he sells it for, add the mark-up to his cost ($42). The mark-up is 110%. To find 110% of his cost, multiply by the decimal equivalent of 110% (1.10); $42 × 1.10 = $46.20. The mark-up is $46.20. Add the mark-up to his cost to find the price the vase sells for; $46.20 + $42.00 = $88.20.

243. c. Use a proportion to solve the problem; $\frac{part}{whole} = \frac{\%}{100}$. The *part* is $125,000 (the part Michelle owns), the *whole* is $400,000 (the whole value of the house), and the *percentage* is x because it is unknown.

$$\frac{125,000}{400,000} = \frac{x}{100}$$

To solve the proportion, cross-multiply, set the cross-products equal to each other, and solve as shown below.

$$\frac{125,000}{400,000} = \frac{x}{100}$$

$$(100)(125,000) = 400,000x$$

$$12,500,000 = 400,000x$$

$$\frac{12,500,000}{400,000} = \frac{400,000x}{400,000}$$

$$x = 31.25$$

Michelle owns 31.25% of the vacation home.

244. d. Find the Social Security tax and the State Disability Insurance, and then subtract the answers from Kyra's weekly wages. To find 7.51% of $895, multiply by the decimal equivalent of 7.51% (0.0751); $895 × 0.0751 = $67.21 (rounded to the nearest cent). Next, find 1.2% of her wages by multiplying by the decimal equivalent of 1.2% (0.012); $895 × 0.012 = $10.74. Subtract $67.21 and $10.74 from Kyra's weekly wages of $895 to find her weekly paycheck; $895 − $67.21 − $10.74 = $817.05. Her weekly paycheck is $817.05.

245. a. Find 5% of the bill by multiplying by the decimal equivalent of 5% (0.05); $178 × 0.05 = $8.90. They will save $8.90.

A common mistake is to use 0.5 instead of 0.05 for 5%; 0.5 is 50%.

246. d. Find 30% of 1,800 by multiplying by the decimal equivalent of 30% (0.30); 1,800 × 0.30 = 540. The maximum number of calories from fats per day is 540.

247. c. Find 24% of $1,345 by multiplying by the decimal equivalent of 24% (0.24); $1,345 × 0.24 = $322.80. $322.80 can be deducted.

248. b. Use the proportion $\frac{part}{whole} = \frac{\%}{100}$. You are looking for the *whole* (100% is the whole capacity of the plant). The *part* you know is 450 and it is 90% of the whole; $\frac{450}{x} = \frac{90}{100}$. To solve the proportion, cross multiply, set the cross-products equal to each other, and solve as shown below.

(450)(100) = 90x

45,000 = 90x

$\frac{45,000}{90} = \frac{90x}{90}$

x = 500

100% capacity is 500 cars.

Another way to look at the problem is to find 10% and multiply it by 10 to get 100%. Given 90%, divide by 9 to find 10%; 450 ÷ 9 = 50. Multiply 10% (50) by 10 to find 100%; 50 × 10 = 500.

249. b. Multiply by the decimal equivalent of $\frac{1}{2}$% (0.005) to find the amount of increase; $152,850 × 0.005 = $764.25. This is how much sales increased. To find the actual amount of sales, add the increase to last month's total; $152,850 + $764.25 = $153,614.25.

A common mistake is to use 0.5 (50%) or 0.05 (5%) for $\frac{1}{2}$%. Rewrite $\frac{1}{2}$% as 0.5%. To find the decimal equivalent, move the decimal point two places to the left. This yields 0.005.

250. **c.** Find 5% of 220 by multiplying 220 by the decimal equivalent of 5% (0.05); 220 × 0.05 = 11 people.

A common mistake is to use 0.5 for 5%; 0.5 is actually 50%.

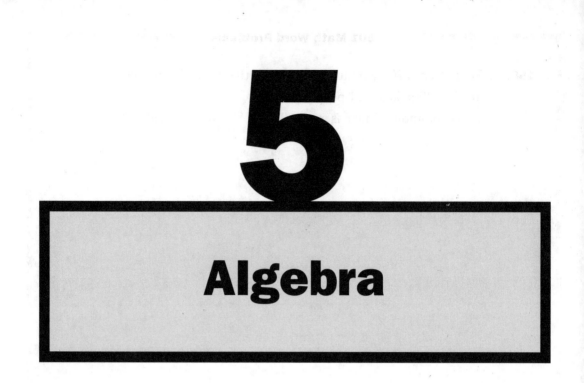

Algebra

Basic algebra problems ask you to solve equations in which one or more elements are unknown. The unknown quantities are represented by variables, which are letters of the alphabet, such as x or y. The questions in this chapter give you practice in writing algebraic equations and using these expressions to solve problems.

251. Assume that the number of hours Katie spent practicing soccer is represented by x. Michael practiced 4 hours less than 2 times the number of hours that Katie practiced. How long did Michael practice?
   a. $2x - 4$
   b. $2x + 4$
   c. $2x + 8$
   d. $4x + 4$

252. Patrick gets paid three dollars less than four times what Kevin gets paid. If the number of dollars that Kevin gets paid is represented by x, what does Patrick get paid?
   a. $3 - 4x$
   b. $3x - 4$
   c. $4x - 3$
   d. $4 - 3x$

253. If the expression $9y - 5$ represents a certain number, which of the following could NOT be the translation?

    a. five less than nine times y

    b. five less than the sum of 9 and y

    c. the difference between $9y$ and 5

    d. the product of nine and y, decreased by 5

254. Susan starts work at 4:00 and Dee starts at 5:00. They both finish at the same time. If Susan works x hours, how many hours does Dee work?

    a. $x + 1$

    b. $x - 1$

    c. x

    d. $2x$

255. Frederick bought 11 books that cost d dollars each. What is the total cost of the books?

    a. $d + 11$

    b. $d - 11$

    c. $11d$

    d. $\frac{d}{11}$

256. There are m months in a year, w weeks in a month and d days in a week. How many days are there in a year?

    a. mwd

    b. $m + w + d$

    c. $\frac{mw}{d}$

    d. $d + \frac{w}{d}$

257. Carlie received x dollars each hour she spent babysitting. She babysat a total of h hours. She then gave half of the money to a friend who had stopped by to help her. How much money did Carlie have after she had paid her friend?

    a. $\frac{hx}{2}$

    b. $\frac{x}{2} + h$

    c. $\frac{h}{2} + x$

    d. $2hx$

258. A long distance call costs x cents for the first minute and y cents for each additional minute. How much would a 10-minute call cost?

  a. $10xy$

  b. $x + 10y$

  c. $\frac{xy}{10}$

  d. $x + 9y$

259. Melissa is four times as old as Jim. Pat is 5 years older than Melissa. If Jim is y years old, how old is Pat?

  a. $4y + 5$

  b. $5y + 4$

  c. $4 \times 5y$

  d. $y + 5$

260. Sally gets paid x dollars per hour for a 40-hour work week and y dollars for each hour she works over 40 hours. How much did Sally earn if she worked 48 hours?

  a. $48xy$

  b. $40y + 8x$

  c. $40x + 8y$

  d. $48x + 48y$

261. Eduardo is combining two 6-inch pieces of wood with a piece that measures 4 inches. How many total inches of wood does he have?

  a. 10 inches

  b. 16 inches

  c. 8 inches

  d. 12 inches

262. Mary has $2 in her pocket. She does yard work for four different neighbors and earns $3 per yard. She then spends $2 on a soda. How much money does she have left?

  a. $18

  b. $10

  c. $12

  d. $14

263. Ten is decreased by four times the quantity of eight minus three. One is then added to that result. What is the final answer?

a. −5

b. −9

c. 31

d. −8

264. The area of a square whose side measures four units is added to the difference of eleven and nine divided by two. What is the total value?

a. 9

b. 16

c. 5

d. 17

265. Four is added to the quantity three minus the sum of negative seven and six. This answer is then multiplied by three. What is the result?

a. 30

b. −21

c. 36

d. 57

266. John and Charlie have a total of 80 dollars. John has x dollars. How much money does Charlie have?

a. 80

b. $80 + x$

c. $80 − x$

d. $x − 80$

267. The temperature in Hillsville was 20° Celsius. What is the equivalent of this temperature in degrees Fahrenheit?

a. 4°

b. 43.1°

c. 68°

d. 132°

268. Peggy's town has an average temperature of 23° Fahrenheit in the winter. What is the average temperature on the Celsius scale?

a. −16.2°

b. 16.2°

c. 5°

d. −5°

269. Celine deposited $505 into her savings account. If the interest rate of the account is 5% per year, how much interest will she have made after 4 years?

a. $252.50

b. $606

c. $10,100

d. $101

270. A certain bank pays 3.4% interest per year for a certificate of deposit, or CD. What is the total balance of an account after 18 months with an initial deposit of $1,250?

a. $765

b. $2,015

c. $63.75

d. $1,313.75

271. Joe took out a car loan for $12,000. He paid $4,800 in interest at a rate of 8% per year. How many years will it take him to pay off the loan?

a. 5

b. 2.5

c. 8

d. 4

272. What is the annual interest rate on an account that earns $711 in simple interest over 36 months with an initial deposit of $7,900?

a. 4.3%

b. 3%

c. 30%

d. 4%

273. Marty used the following mathematical statement to show he could change an expression and still get the same answer on both sides:
$$10 \times (6 \times 5) = (10 \times 6) \times 5$$
Which mathematical property did Marty use?

a. Identity Property of Multiplication

b. Commutative Property of Multiplication

c. Distributive Property of Multiplication over Addition

d. Associative Property of Multiplication

274. Tori was asked to give an example of the commutative property of addition. Which of the following choices would be correct?

 a. $3 + (4 + 6) = (3 + 4) + 6$

 b. $3(4 + 6) = 3(4) + 3(6)$

 c. $3 + 4 = 4 + 3$

 d. $3 + 0 = 3$

275. Jake needed to find the perimeter of an equilateral triangle whose sides measure $x + 4$ cm each. Jake realized that he could multiply $3 (x + 4) = 3x + 12$ to find the total perimeter in terms of x. Which property did he use to multiply?

 a. Associative Property of Addition

 b. Distributive Property of Multiplication over Addition

 c. Commutative Property of Multiplication

 d. Inverse Property of Addition

276. The product of -5 and a number is 35. What is the number?

 a. 40

 b. 30

 c. -7

 d. -35

277. When ten is subtracted from the opposite of a number, the difference between them is five. What is the number?

 a. 15

 b. -15

 c. -5

 d. 5

278. The sum of -4 and a number is equal to -48. What is the number?

 a. -12

 b. -44

 c. 12

 d. -52

279. Twice a number increased by 11 is equal to 32 less than three times the number. Find the number.

 a. -21

 b. $\frac{21}{5}$

 c. 43

 d. $\frac{43}{5}$

280. If one is added to the difference when $10x$ is subtracted from $-18x$, the result is 57. What is the value of x?

    a. -2

    b. -7

    c. 2

    d. 7

281. If 0.3 is added to 0.2 times the quantity $x - 3$, the result is 2.5. What is the value of x?

    a. 1.7

    b. 26

    c. 14

    d. 17

282. If twice the quantity $x + 6$ is divided by negative seven, the result is 6. Find the number.

    a. -18

    b. -27

    c. -15

    d. -0.5

283. The difference between six times the quantity $6x + 1$ and three times the quantity $x - 1$ is 108. What is the value of x?

    a. $\frac{12}{11}$

    b. $\frac{35}{11}$

    c. 12

    d. 3

284. Negative four is multiplied by the quantity $x + 8$. If $6x$ is then added to this, the result is $2x + 32$. What is the value of x?

    a. No solution

    b. Identity

    d. 0

    d. 16

285. Patrice has worked a certain amount of hours so far this week. Tomorrow she will work four more hours to finish out the week with a total of 10 hours. How many hours has she worked so far?

    a. 40

    b. 14

    c. 6

    d. 2.5

286. Michael has 16 CDs. This is four more than one half the amount that Kathleen has. How many CDs does Kathleen have?

    a. 6

    b. 24

    d. 4

    d. 12

287. The perimeter of a square can be expressed as $x + 4$. If one side of the square is 24, what is the value of x?

    a. 2

    b. 7

    c. 5

    d. 92

288. The perimeter of a rectangle is 21 inches. What is the measure of its width if its length is 3 inches greater than its width?

    a. 9

    b. 3.75

    c. 4.5

    d. 3

289. The sum of two consecutive integers is 41. What are the integers?

    a. 20, 21

    b. 21, 22

    c. 20, 22

    d. 10.5, 10.5

290. The sum of two consecutive even integers is 126. What are the integers?

    a. 62, 64

    b. 62, 63

    b. 60, 66

    d. 2, 63

291. The sum of two consecutive odd integers is –112. What is the larger integer?

    a. –55

    b. –57

    c. 55

    d. 57

292. The sum of three consecutive even integers is 114. What is the value of the largest consecutive integer?

    a. 36

    b. 34

    c. 40

    d. 38

293. Two commuters leave the same city at the same time but travel in opposite directions. One car is traveling at an average speed of 63 miles per hour, and the other car is traveling at an average speed of 59 miles per hour. How many hours will it take before the cars are 610 miles apart?

    a. 4

    b. 6

    c. 30

    d. 5

294. Two trains leave the same city at the same time, one going east and the other going west. If one train is traveling at 65 mph and the other at 72 mph, how many hours will it take for them to be 822 miles apart?

    a. 9

    b. 7

    c. 8

    d. 6

295. Two trains leave two different cities 1,029 miles apart and head directly toward each other on parallel tracks. If one train is traveling at 45 miles per hour and the other at 53 miles per hour, how many hours will it take before the trains pass?

    a. 9.5

    b. 11

    c. 11.5

    d. 10.5

296. Nine minus five times a number, x, is no less than 39. Which of the following expressions represents all the possible values of the number?

    a. $x \le 6$

    b. $x \ge -6$

    c. $x \le -6$

    d. $x \ge 6$

297. Will has a bag of gumdrops. If he eats 2 of his gumdrops, he will have between 2 and 6 of them left. Which of the following represents how many gumdrops, x, were originally in his bag?

    a. $4 < x < 8$

    b. $0 < x < 4$

    c. $0 > x > 4$

    d. $4 > x > 8$

298. The value of y is between negative three and positive eight inclusive. Which of the following represents y?

    a. $-3 \le y \le 8$

    b. $-3 < y \le 8$

    c. $-3 \le y < 8$

    d. $-3 \ge y \ge 8$

299. Five more than the quotient of a number and 2 is at least that number. What is the greatest value of the number?

    a. 7

    b. 10

    c. 5

    d. 2

300. Carl worked three more than twice as many hours as Cindy did. What is the maximum amount of hours Cindy worked if together they worked 48 hours at most?

    a. 17

    b. 33

    c. 37

    d. 15

301. The cost of renting a bike at the local bike shop can be represented by the equation $y = 2x + 2$, where y is the total cost and x is the number of hours the bike is rented. Which of the following ordered pairs would be a possible number of hours rented, x, and the corresponding total cost, y?

    a. $(0, -2)$
    b. $(2, 6)$
    c. $(6, 2)$
    d. $(-2, -6)$

302. A telephone company charges $.35 for the first minute of a phone call and $.15 for each additional minute of the call. Which of the following represents the cost y of a phone call lasting x minutes?

    a. $y = 0.15(x - 1) + 0.35$
    b. $x = 0.15(y - 1) + 0.35$
    c. $y = 0.15x + 0.35$
    d. $x = 0.15y + 0.35$

303. A ride in a taxicab costs $1.25 for the first mile and $1.15 for each additional mile. Which of the following could be used to calculate the total cost y of a ride that was x miles?

    a. $x = 1.25(y - 1) + 1.15$
    b. $x = 1.15(y - 1) + 1.25$
    c. $y = 1.25(x - 1) + 1.15$
    d. $y = 1.15(x - 1) + 1.25$

304. The cost of shipping a package through Shipping Express is $4.85 plus $2 per ounce of the weight of the package. Sally only has $10 to spend on shipping costs. Which of the following could Sally use to find the maximum number of ounces she can ship for $10?

    a. $4.85x + 2 \leq 10$
    b. $4.85x + 2 \geq 10$
    c. $2x + 4.85 \leq 10$
    d. $2x + 4.85 \geq 10$

305. Green Bank charges a monthly fee of $3 for a checking account and $.10 per check. Savings-R-Us bank charges a $4.50 monthly fee and $.05 per check. How many checks need to be used for the monthly costs to be the same for both banks?

  a. 25
  b. 30
  c. 35
  d. 100

306. Easy Rider taxi service charges a pick-up fee of $2 and $1.25 for each mile. Luxury Limo taxi service charges a pick-up fee of $3.25 and $1 per mile. How many miles need to be driven for both services to cost the same amount?

  a. 24
  b. 12
  c. 10
  d. 5

307. The sum of two integers is 36, and the difference is 6. What is the smaller of the two numbers?

  a. 21
  b. 15
  c. 16
  d. 18

308. One integer is two more than another. The sum of the lesser integer and twice the greater is 7. What is the greater integer?

  a. 1
  b. 2
  c. 3
  d. 7

309. One integer is four times another. The sum of the integers is 65. What is the value of the lesser integer?

  a. 42
  b. 64
  c. 30
  d. 13

310. The sum of three times a greater integer and 5 times a lesser integer is 9. Three less than the greater equals the lesser. What is the value of the lesser integer?

    a. 0

    b. 1

    c. 2

    d. 3

311. The perimeter of a rectangle is 104 inches. The width is 6 inches less than 3 times the length. Find the width of the rectangle.

    a. 13.5 inches

    b. 37.5 inches

    c. 14.5 inches

    d. 15 inches

312. The perimeter of a parallelogram is 50 cm. The length of the parallelogram is 5 cm more than the width. Find the length of the parallelogram.

    a. 15 cm

    b. 11 cm

    c. 5 cm

    d. 10 cm

313. Jackie invested money in two different accounts, one of which earned 12% interest per year and another that earned 15% interest per year. The amount invested at 15% was 100 more than twice the amount at 12%. How much was invested at 12% if the total annual interest earned was $855?

    a. $4,100

    b. $2,100

    c. $2,000

    d. $4,000

314. Kevin invested $4,000 in an account that earns 6% interest per year and $x in a different account that earns 8% interest per year. How much is invested at 8% if the total amount of interest earned annually is $405.50?
   a. $2,075.00
   b. $4,000.00
   c. $2,068.75
   d. $2,075.68

315. Megan bought x pounds of coffee that cost $3 per pound and 18 pounds of coffee at $2.50 per pound for the company picnic. Find the total number of pounds of coffee purchased if the average cost per pound of both types together is $2.85.
   a. 42
   b. 18
   c. 63
   d. 60

316. The student council bought two different types of candy for the school fair. They purchased 40 pounds of candy at $2.15 per pound and x pounds at $1.90 per pound. What is the total number of pounds they bought if the total amount of money spent on candy was $162?
   a. 42
   b. 40
   c. 80
   d. 52

317. The manager of a garden store ordered two different kinds of marigold seeds for her display. The first type cost her $1 per packet and the second type cost $1.26 per packet. How many packets of the first type did she purchase if she bought 50 more of the $1.26 packets than the $1 packets and spent a total of $402?
   a. 150
   b. 200
   c. 250
   d. 100

318. Harold used a 3% iodine solution and a 20% iodine solution to make a 95-ounce solution that was 19% iodine. How many ounces of the 3% iodine solution did he use?

a. 5

b. 80

c. 60

d. 20

319. A chemist mixed a solution that was 34% acid with another solution that was 18% acid to produce a 30-ounce solution that was 28% acid. How much of the 34% acid solution did he use?

a. 27

b. 11.25

c. 18.75

d. 28

320. Bob is 2 years from being twice as old as Ellen. The sum of twice Bob's age and three times Ellen's age is 66. How old is Ellen?

a. 15

b. 10

c. 18

d. 20

321. Sam's age is 1 less than twice Shari's age. The sum of their ages is 92. How old is Shari?

a. 48

b. 32

c. 65

d. 31

322. At the school bookstore, two binders and three pens cost $12.50. Three binders and five pens cost $19.50. What is the total cost of 1 binder and 1 pen?

a. $4.50

b. $4.00

c. $1.50

d. $5.50

323. Two angles are complementary. The larger angle is 15° more than twice the smaller. Find the measure of the smaller angle.

a. 25°

b. 65°

c. 90°

d. 82.5°

324. The cost of a student ticket is $1 more than half of an adult ticket. Six adults and four student tickets cost $76. What is the cost of one adult ticket?

a. $2.50

b. $3.00

c. $5.50

d. $4.00

325. Three shirts and five ties cost $23. Five shirts and one tie cost $20. What is the price of one shirt?

a. $3.50

b. $2.50

c. $6.00

d. $3.00

326. Noel rode $3x$ miles on his bike and Jamie rode $5x$ miles on hers. In terms of x, what is the total number of miles they rode?

a. $15x$ miles

b. $15x^2$ miles

c. $8x$ miles

d. $8x^2$ miles

327. If the areas of two sections of a garden are $6a + 2$ and $5a$, what is the difference between the areas of the two sections in terms of a?

a. $a - 2$

b. $3a + 2$

c. $a + 2$

d. $11a - 2$

328. Laura has a rectangular garden whose width is x^3 and whose length is x^4. In terms of x, what is the area of her garden?

a. $2x^7$

b. x^7

c. x^{12}

d. $2x^{12}$

329. Jonestown High School has a soccer field whose dimensions can be expressed as $7y^2$ and $3xy$. What is the area of this field in terms of x and y?

a. $10xy^2$

b. $10xy^3$

c. $21xy^3$

d. $21xy^2$

330. The area of a parallelogram is x^8. If the base is x^4, what is the height in terms of x?

a. x^4

b. x^2

c. x^{12}

d. x^{32}

331. The quotient of $3d^4$ and $12d^6$ is

a. $4d^2$.

b. $4d^8$.

c. $\frac{1}{4d^8}$.

d. $\frac{1}{4d^2}$.

332. The product of $6x^2$ and $4xy^2$ is divided by $3x^3y$. What is the simplified expression?

a. $8y$

b. $\frac{4y}{x}$

c. $4y$

d. $\frac{8y}{x}$

333. If the side of a square can be expressed as a^2b^3, what is the area of the square in simplified form?

a. a^4b^5

b. a^4b^6

c. a^2b^6

d. a^2b^5

334. If $3x^2$ is multiplied by the quantity $2x^3y$ raised to the fourth power, what would this expression simplify to?

 a. $48x^{14}y^4$
 b. $1,296x^{16}y^4$
 c. $6x^9y^4$
 d. $6x^{14}y^4$

335. Sara's bedroom is in the shape of a rectangle. The dimensions are $2x$ and $4x + 5$. What is the area of Sara's bedroom?

 a. $18x$
 b. $18x^2$
 c. $8x^2 + 5x$
 d. $8x^2 + 10x$

336. Express the product of $-9p^3r$ and the quantity $5p - 6r$ in simplified form.

 a. $-4p^4r - 15p^3r^2$
 b. $-45p^4r + 54p^3r^2$
 c. $-45p^4r - 6r$
 d. $-45p^3r + 54p^3r^2$

337. A number, x, increased by 3 is multiplied by the same number, x, increased by 4. What is the product of the two numbers in terms of x?

 a. $x^2 + 7$
 b. $x^2 + 12$
 c. $x^2 + 7x + 12$
 d. $x^2 + x + 7$

338. The length of Kara's rectangular patio can be expressed as $2x - 1$ and the width can be expressed as $x + 6$. In terms of x, what is the area of her patio?

 a. $2x^2 + 13x - 6$
 b. $2x^2 - 6$
 c. $2x^2 - 5x - 6$
 d. $2x^2 + 11x - 6$

339. A car travels at a rate of $(4x^2 - 2)$. What is the distance this car will travel in $(3x - 8)$ hours?

 a. $12x^3 - 32x^2 - 6x + 16$
 b. $12x^2 - 32x^2 - 6x + 16$
 c. $12x^3 + 32x^2 - 6x - 16$
 d. $12x^3 - 32x^2 - 5x + 16$

340. The area of the base of a prism can be expressed as $x^2 + 4x + 1$ and the height of the prism can be expressed as $x - 3$. What is the volume of this prism in terms of x?

a. $x^3 + x^2 - 13x - 3$

b. $x^3 + 7x^2 - 13x - 3$

c. $x^3 - x^2 - 11x - 3$

d. $x^3 + x^2 - 11x - 3$

341. The dimensions of a rectangular prism can be expressed as $x + 1$, $x - 2$, and $x + 4$. In terms of x, what is the volume of the prism?

a. $x^3 + 3x^2 + 6x - 8$

b. $x^3 + 3x^2 - 6x - 8$

c. $x^3 + 5x^2 - 2x + 8$

d. $x^3 - 5x^2 - 2x - 8$

342. The area of Mr. Smith's rectangular classroom is $x^2 - 36$. Which of the following binomials could represent the length and the width of the room?

a. $(x + 6)(x + 6)$

b. $(x - 6)(x - 6)$

c. $(x + 6)(x - 6)$

d. $x(x - 36)$

343. The area of a parallelogram can be expressed as the binomial $2x^2 - 10x$. Which of the following could be the length of the base and the height of the parallelogram?

a. $2x(x^2 - 5x)$

b. $2x (x - 5)$

c. $(2x - 1)(x - 10)$

d. $(2x - 5)(x + 2)$

344. A farmer's rectangular field has an area that can be expressed as the trinomial $x^2 + 2x + 1$. In terms of x, what are the dimensions of the field?

a. $(x + 1)(x + 2)$

b. $(x - 1)(x - 2)$

c. $(x - 1)(x + 2)$

d. $(x + 1)(x + 1)$

345. Harold is tiling a rectangular kitchen floor with an area that is expressed as $x^2 + 6x + 5$. What could the dimensions of the floor be in terms of x?

a. $(x + 1)(x + 5)$

b. $(x - 1)(x - 5)$

c. $(x - 2)(x + 3)$

d. $(x + 2)(x + 3)$

346. The area of a rectangle is represented by the trinomial: $x^2 + x - 12$. Which of the following binomials could represent the length and width?

a. $(x + 4)(x - 3)$

b. $(x - 4)(x - 3)$

c. $(x - 4)(x + 3)$

d. $(x - 6)(x + 2)$

347. Katie's school has a rectangular courtyard whose area can be expressed as $3x^2 - 7x + 2$. Which of the following could be the dimensions of the courtyard in terms of x?

a. $(3x - 1)(x + 2)$

b. $(3x - 1)(x - 2)$

c. $(3x - 2)(x - 1)$

d. $(3x + 2)(x + 1)$

348. The distance from the sun to the earth is approximately 9.3×10^7 miles. What is this distance expressed in standard notation?

a. 930,000,000

b. 93,700,000

c. 0.00000093

d. 93,000,000

349. The distance from the earth to the moon is approximately 240,000 miles. What is this distance expressed in scientific notation?

a. 24×10^4

b. 240×10^3

c. 2.4×10^5

d. 2.4×10^{-5}

350. It takes light 5.3×10^{-6} seconds to travel one mile. What is this time in standard notation?

   a. 0.00000053

   b. 0.000053

   c. 5.300000

   d. 0.0000053

351. The square of a positive number is 49. What is the number?

   a. $\sqrt{7}$

   b. −7

   c. 7 or −7

   d. 7

352. The square of a number added to 25 equals 10 times the number. What is the number?

   a. −5

   b. 10

   c. −10

   d. 5

353. The sum of the square of a number and 12 times the number is −27. What is the smaller possible value of this number?

   a. −3

   b. −9

   c. 3

   d. 9

354. The area of a rectangle is 24 square inches. The length of the rectangle is 5 inches more than the width. How many inches is the width?

   a. 4 in

   b. 3 in

   c. 6 in

   d. 8 in

355. The height of a parallelogram measures 5 meters more than its base. If the area of the parallelogram is 36 m^2, what is the height in meters?

   a. 6 m

   b. 9 m

   c. 12 m

   d. 4 m

356. Patrick has a rectangular patio whose length is 5 m less than the diagonal and a width that is 7 m less than the diagonal. If the area of his patio is 195 m², what is the length of the diagonal?

    a. 10 m

    b. 8 m

    c. 16 m

    d. 20 m

357. Samantha owns a rectangular field that has an area of 3,280 square feet. The length of the field is 2 more than twice the width. What is the width of the field?

    a. 40 ft

    b. 82 ft

    c. 41 ft

    d. 84 ft

358. A garden in the shape of a rectangle is surrounded by a walkway of uniform width. The dimensions of the garden only are 35 by 24. The area of the garden and the walkway together is 1,530 square feet. What is the width of the walkway in feet?

    a. 4 ft

    b. 5 ft

    c. 34.5 ft

    d. 24 ft

359. A pool is surrounded by a deck that has the same width all the way around. The total area of the deck only is 400 square feet. The dimensions of the pool are 18 feet by 24 feet. How many feet is the width of the deck?

    a. 4 ft

    b. 8 ft

    c. 24 ft

    d. 25 ft

360. Jessica has a picture in a frame with a total area of 288 in². The dimension of the picture without the frame is 12 in by 14 in. What is the larger dimension, in inches, of the frame?

    a. 2 in

    b. 14 in

    c. 18 in

    d. 16 in

361. What is the lesser of two consecutive positive integers whose product is 90?

   a. –9
   b. 9
   c. –10
   d. 10

362. What is the greater of two consecutive negative integers whose product is 210?

   a. –14
   b. –15
   c. 14
   d. 15

363. Find the lesser of two consecutive positive even integers whose product is 168.

   a. 12
   b. 14
   c. 10
   d. 16

364. Find the greater of two consecutive positive odd integers whose product is 143.

   a. 10
   b. 11
   c. 12
   d. 13

365. The sum of the squares of two consecutive positive odd integers is 74. What is the value of the smaller integer?

   a. 3
   b. 7
   c. 5
   d. 11

366. If the difference between the squares of two consecutive integers is 15, find the larger integer.

   a. 8
   b. 7
   c. 6
   d. 9

367. The square of one integer is 55 less than the square of the next consecutive integer. Find the lesser integer.

    a. 23

    b. 24

    c. 27

    d. 28

368. A 4-inch by 6-inch photograph is going to be enlarged by increasing each side by the same amount. The new area is 168 square inches. How many inches is each dimension increased?

    a. 12

    b. 10

    c. 8

    d. 6

369. A photographer decides to reduce a picture she took in order to fit it into a certain frame. She needs the picture to be one-third of the area of the original. If the original picture was 4 inches by 6 inches, how many inches is the smaller dimension of the reduced picture if each dimension changes the same amount?

    a. 2

    b. 3

    c. 4

    d. 5

370. A rectangular garden has a width of 20 feet and a length of 24 feet. If each side of the garden is increased by the same amount, how many feet is the new length if the new area is 141 square feet more than the original?

    a. 23

    b. 24

    c. 26

    d. 27

371. Ian can remodel a kitchen in 20 hours and Jack can do the same job in 15 hours. If they work together, how many hours will it take them to remodel the kitchen?

    a. 5.6

    b. 8.6

    c. 7.5

    d. 12

372. Peter can paint a room in an hour and a half and Joe can paint the same room in 2 hours. How many minutes will it take them to paint the room if they do it together? Round answer to nearest minute.

   a. 51

   b. 64

   c. 30

   d. 210

373. Carla can plant a garden in 3 hours and Charles can plant the same garden in 4.5 hours. If they work together, how many hours will it take them to plant the garden?

   a. 1.5

   b. 2.1

   c. 1.8

   d. 7.5

374. If Jim and Jerry work together they can finish a job in 4 hours. If working alone takes Jim 10 hours to finish the job, how many hours would it take Jerry to do the job alone?

   a. 16

   b. 5.6

   c. 6.7

   d. 6.0

375. Bill and Ben can clean the garage together in 6 hours. If it takes Bill 15 hours working alone, how long will it take Ben working alone?

   a. 11 hours

   b. 9 hours

   c. 16 hours

   d. 10 hours

Answer Explanations

The following explanations show one way in which each problem can be solved. You may have another method for solving these problems.

251. a. The translation of "two times the number of hours" is $2x$. Four hours less than $2x$ becomes $2x - 4$.

252. c. When the key words *less than* appear in a sentence, it means that you will subtract from the next part of the sentence, so it will appear at the end of the expression. "Four times a number" is equal to $4x$ in this problem. Three less than $4x$ is $4x - 3$.

253. b. Each one of the answer choices would translate to $9y - 5$ except for choice **b.** The word *sum* is a key word for addition, and $9y$ means "9 times y."

254. b. Since Susan started 1 hour before Dee, Dee has been working for one less hour than Susan had been working. Thus, $x - 1$.

255. c. Frederick would multiply the number of books, 11, by how much each one costs, d. For example, if each one of the books cost $10, he would multiply 11 times $10 and get $110. Therefore, the answer is $11d$.

256. a. In this problem, multiply d and w to get the total days in one month and then multiply that result by m, to get the total days in the year. This can be expressed as mwd, which means m times w times d.

257. a. To calculate the total she received, multiply x dollars per hour times h, the number of hours she worked. This becomes xh. Divide this amount by 2 since she gave half to her friend. Thus, $\frac{xh}{2}$ is how much money she has left.

258. d. The cost of the call is x cents plus y times the additional minutes. Since the call is 10 minutes long, she will pay x cents for 1 minute and y cents for the other nine. Therefore the expression is $1x + 9y$, or $x + 9y$, since it is not necessary to write a 1 in front of a variable.

259. a. Start with Jim's age, y, since he appears to be the youngest. Melissa is four times as old as he is, so her age is $4y$. Pat is 5 years older than Melissa, so Pat's age would be Melissa's age, $4y$, plus another 5 years. Thus, $4y + 5$.

260. c. Since she worked 48 hours, Sally will get paid her regular amount, x dollars, for 40 hours and a different amount, y, for the additional 8 hours. This becomes 40 *times* x plus 8 *times* y, which translates to $40x + 8y$.

261. b. This problem translates to the expression $6 \times 2 + 4$. Using order of operations, do the multiplication first; $6 \times 2 = 12$ and then add $12 + 4 = 16$ inches.

262. c. This translates to the expression $2 + 3 \times 4 - 2$. Using order of operations, multiply 3×4 first; $2 + 12 - 2$. Add and subtract the numbers in order from left to right; $2 + 12 = 14$; $14 - 2 = 12$.

263. b. This problem translates to the expression $10 - 4(8 - 3) + 1$. Using order of operations, do the operation inside the parentheses first; $10 - 4(5) + 1$. Since multiplication is next, multiply 4×5; $10 - 20 + 1$. Add and subtract in order from left to right; $10 - 20 = -10$; $-10 + 1 = -9$.

264. d. This problem translates to the expression $4^2 + (11 - 9) \div 2$. Using order of operations, do the operation inside the parentheses first; $4^2 + (2) \div 2$. Evaluate the exponent; $16 + (2) \div 2$. Divide $2 \div 2$; $16 + 1$. Add; $16 + 1 = 17$.

265. c. This problem translates to the expression $3\{[3 - (-7 + 6)] + 8\}$. When dealing with multiple grouping symbols, start from the innermost set and work your way out. Add and subtract in order from left to right inside the brackets. Remember that subtraction is the same as adding the opposite so $3 - (-1)$ becomes $3 + (+1) = 4$; $3\{[3 - (-1)] + 8\}$; $3[4 + 8]$. Multiply 3×12 to finish the problem; $3[12] = 36$.

266. c. If the total amount for both is 80, then the amount for one person is 80 minus the amount of the other person. Since John has x dollars, Charlie's amount is $80 - x$.

267. c. Use the formula $F = \frac{9}{5}C + 32$. Substitute the Celsius temperature of 20° for C in the formula. This results in the equation $F = \frac{9}{5}(20) + 32$. Following the order of operations, multiply $\frac{9}{5}$ and 20 to get 36. The final step is to add $36 + 32$ for an answer of 68°.

268. d. Use the formula $C = \frac{5}{9}(F - 32)$. Substitute the Fahrenheit temperature of 23° for F in the formula. This results in the equation $C = \frac{5}{9}(23 - 32)$. Following the order of operations, begin calculations inside the parentheses first and subtract $23 - 32$ to get -9. Multiply $\frac{5}{9}$ times -9 to get an answer of $-5°$.

269. d. Using the simple interest formula *Interest = principal × rate × time*, or $I = prt$, substitute $p = \$505$, $r = .05$ (the interest rate as a decimal) and $t = 4$; $I = (505)(.05)(4)$. Multiply to get a result of $I = \$101$.

270. d. Using the simple interest formula *Interest = principal × rate × time*, or $I = prt$, substitute $p = \$1,250$, $r = 0.034$ (the interest rate as a decimal), and $t = 1.5$ (18 months is equal to 1.5 years); $I = (1,250)(.034)(1.5)$. Multiply to get a result of $I = \$63.75$. To find the total amount in the account after 18 months, add the interest to the initial principal. $\$63.75 + \$1,250 = \$1,313.75$.

271. a. Using the simple interest formula *Interest = principal × rate × time*, or $I = prt$, substitute $I = \$4,800$, $p = \$12,000$, and $r = .08$ (the interest rate as a decimal); $4,800 = (12,000)(.08)(t)$. Multiply 12,000 and .08 to get 960, so $4,800 = 960t$. Divide both sides by 960 to get $5 = t$. Therefore, the time is 5 years.

272. b. Using the simple interest formula *Interest = principal × rate × time*, or $I = prt$, substitute $I = \$711$, $p = \$7,900$, and $t = 3$ (36 months is equal to 3 years); $711 = (7,900)(r)(3)$. Multiply 7,900 and 3 on the right side to get a result of $711 = 23,700r$. Divide both sides by 23,700 to get $r = .03$, which is a decimal equal to 3%.

273. d. In the statement, the order of the numbers does not change; however, the grouping of the numbers in parentheses does. Each side, if simplified, results in an answer of 300, even though both sides look different. Changing the grouping in a problem like this is an example of the associative property of multiplication.

274. c. Choice **a** is an example of the associative property of addition, where changing the grouping of the numbers will still result in the same answer. Choice **b** is an example of the distributive property of multiplication over addition. Choice **d** is an example of the additive identity, where any number added to zero equals itself. Choice **c** is an example of the commutative property of addition, where we can change the order of the numbers that are being added and the result is always the same.

275. b. In the statement, 3 is being multiplied by the quantity in the parentheses, $x + 4$. The distributive property allows you to multiply $3 \times x$ and add it to 3×4, simplifying to $3x + 12$.

276. c. Let $y =$ the number. The word *product* is a key word for multiplication. Therefore the equation is $-5y = 35$. To solve this, divide each side of the equation by -5; $\frac{-5y}{-5} = \frac{35}{-5}$. The variable is now alone: $y = -7$.

277. b. Let $x =$ the number. The opposite of this number is $-x$. The words *subtraction* and *difference* both tell you to subtract, so the equation becomes $-x - 10 = 5$. To solve this, add 10 to both sides of the equation; $-x - 10 + 10 = 5 + 10$. Simplify to $-x = 15$. Divide both sides of the equation by -1. Remember that $-x = -1x$; $\frac{-x}{-1} = \frac{15}{-1}$. The variable is now alone: $x = -15$.

278. b. Let $x =$ the number. Since *sum* is a key word for addition, the equation is $-4 + x = -48$. Add 4 to both sides of the equation; $-4 + 4 + x = -48 + 4$. The variable is now alone: $x = -44$.

279. c. Let $x =$ the number. Now translate each part of the sentence. Twice a number increased by 11 is $2x + 11$; 32 less than 3 times a number is $3x - 32$. Set the expressions equal to each other: $2x + 11 = 3x - 32$. Subtract $2x$ from both sides of the equation: $2x - 2x + 11 = 3x - 2x - 32$. Simplify: $11 = x - 32$. Add 32 to both sides of the equation: $11 + 32 = x - 32 + 32$. The variable is now alone: $x = 43$.

280. a. The statement, "If one is added to the difference when $10x$ subtracted from $-18x$, the result is 57," translates to the equation $-18x - 10x + 1 = 57$. Combine like terms on the left side of the equation: $-28x + 1 = 57$. Subtract 1 from both sides of the equation: $-28x + 1 - 1 = 57 - 1$. Divide each side of the equation by -28: $\frac{-28x}{-28} = \frac{56}{-28}$. The variable is now alone: $x = -2$.

281. c. The statement, "If 0.3 is added to 0.2 times the quantity $x - 3$, the result is 2.5," translates to the equation $0.2(x - 3) + 0.3 = 2.5$. Remember to use parentheses for the expression when the words *the quantity* are used. Use the distributive property on the left side of the equation: $0.2x - 0.6$

+ 0.3 = 2.5. Combine like terms on the left side of the equation: $0.2x + -0.3 = 2.5$. Add 0.3 to both sides of the equation: $0.2x + -0.3 + 0.3 = 2.5 + 0.3$. Simplify: $0.2x = 2.8$. Divide both sides by 0.2: $\frac{0.2x}{0.2} = \frac{2.8}{0.2}$. The variable is now alone: $x = 14$.

282. b. Let x = the number. The sentence, "If twice the quantity $x + 6$ is divided by negative seven, the result is 6," translates to $\frac{2(x + 6)}{-7} = 6$. Remember to use parentheses for the expression when the words *the quantity* are used.

   There are different ways to approach solving this problem.

Method I:
Multiply both sides of the equation by -7: $-7 \times \frac{2(x + 6)}{-7} = 6 \times -7$
This simplifies to: $2(x + 6) = -42$
Divide each side of the equation by 2: $\frac{2(x + 6)}{2} = \frac{-42}{2}$
This simplifies to: $x + 6 = -21$
Subtract 6 from both sides of the equation: $x + 6 - 6 = -21 - 6$
The variable is now alone: $x = -27$

Method II:
Another way to look at the problem is to multiply each side by -7 in the first step to get: $2(x + 6) = -42$
Then use distributive property on the left side: $2x + 12 = -42$
Subtract 12 from both sides of the equation: $2x + 12 - 12 = -42 - 12$
Simplify: $2x = -54$
Divide each side by 2: $\frac{2x}{2} = \frac{-54}{2}$
The variable is now alone: $x = -27$

283. d. Translating the sentence, "The difference between six times the quantity $6x + 1$ minus three times the quantity $x - 1$ is 108," into symbolic form results in the equation: $6(6x + 1) - 3(x - 1) = 108$. Remember to use parentheses for the expression when the words *the quantity* are used. Perform the distributive property twice on the left side of the equation: $36x + 6 - 3x + 3 = 108$. Combine like terms on the left side of the equation: $33x + 9 = 108$. Subtract 9 from both sides of the equation: $33x + 9 - 9 = 108 - 9$. Simplify: $33x = 99$. Divide both sides of the equation by 33: $\frac{33x}{33} = \frac{99}{33}$. The variable is now alone: $x = 3$.

284. a. This problem translates to the equation $-4(x + 8) + 6x = 2x + 32$. Remember to use parentheses for the expression when the words *the quantity* are used. Use distributive property on the left side of the equation: $-4x - 32 + 6x = 2x + 32$. Combine like terms on the left side of the equation: $2x - 32 = 2x + 32$. Subtract $2x$ from both sides of the equation: $2x - 2x - 32 = 2x - 2x + 32$. The two sides are not equal. There is no solution: $-32 \neq 32$.

285. c. Let x = the amount of hours worked so far this week. Therefore, the equation is $x + 4 = 10$. To solve this equation, subtract 4 from both sides of the equation; $x + 4 - 4 = 10 - 4$. The variable is now alone: $x = 6$.

286. b. Let x = the number of CDs Kathleen has. Four more than one half the number can be written as $\frac{1}{2}x + 4$. Set this amount equal to 16, which is the number of CDs Michael has. To solve this, subtract 4 from both sides of the equation: $\frac{1}{2}x + 4 - 4 = 16 - 4$. Multiply each side of the equation by 2: $\frac{2x}{2} = 2 \times 12$. The variable is now alone: $x = 24$.

287. d. Since the perimeter of the square is $x + 4$, and a square has four equal sides, we can use the perimeter formula for a square to find the answer to the question: $P = 4s$ where P = perimeter and s = side length of the square. Substituting the information given in the problem, $P = x + 4$ and $s = 24$, gives the equation: $x + 4 = 4(24)$. Simplifying yields $x + 4 = 96$. Subtract 4 from both sides of the equation: $x + 4 - 4 = 96 - 4$. Simplify: $x = 92$.

288. b. Let x = the width of the rectangle. Let $x + 3$ = the length of the rectangle, since the length is "3 more than" the width. Perimeter is the distance around the rectangle. The formula is length + width + length + width, $P = l + w + l + w$, or $P = 2l + 2w$. Substitute the *let* statements for l and w and the perimeter (P) equal to 21 into the formula: $21 = 2(x + 3) + 2(x)$. Use the distributive property on the right side of the equation: $21 = 2x + 6 + 2x$. Combine like terms of the right side of the equation: $21 = 4x + 6$. Subtract 6 from both sides of the equation: $21 - 6 = 4x + 6 - 6$. Simplify: $15 = 4x$. Divide both sides of the equation by 4: $\frac{15}{4} = \frac{4x}{4}$. The variable is now alone: $3.75 = x$.

289. a. Two consecutive integers are numbers in order like 4 and 5 or -30 and -29, which are each 1 number apart. Let x = the first consecutive integer. Let $x + 1$ = the second consecutive integer. *Sum* is a key word for addition so the equation becomes: $(x) + (x + 1) = 41$. Combine like terms on the left side of the equation: $2x + 1 = 41$. Subtract 1 from both sides of the equation: $2x + 1 - 1 = 41 - 1$. Simplify: $2x = 40$. Divide each side of the equation by 2: $\frac{2x}{2} = \frac{40}{2}$. The variable is now alone: $x = 20$. Therefore the larger integer is: $x + 1 = 21$. The two integers are 20 and 21.

290. a. Two consecutive **even** integers are numbers in order, such as 4 and 6 or –30 and –32, which are each 2 numbers apart. Let x = the first consecutive even integer. Let $x + 2$ = the second (and larger) consecutive even integer. *Sum* is a key word for addition so the equation becomes $(x) + (x + 2) = 126$. Combine like terms on the left side of the equation: $2x + 2 = 126$. Subtract 2 from both sides of the equation: $2x + 2 - 2 = 126 - 2$; simplify: $2x = 124$. Divide each side of the equation by 2: $\frac{2x}{2} = \frac{124}{2}$. The variable is now alone: $x = 62$. Therefore the larger integer is: $x + 2 = 64$.

291. a. Two consecutive **odd** integers are numbers in order like 3 and 5 or –31 and –29, which are each 2 numbers apart. In this problem you are looking for 2 consecutive odd integers. Let x = the first and smallest consecutive odd integer. Let $x + 2$ = the second (and larger) consecutive negative odd integer. *Sum* is a key word for addition so the equation becomes $(x) + (x + 2) = -112$. Combine like terms on the left side of the equation: $2x + 2 = -112$. Subtract 2 from both sides of the equation: $2x + 2 - 2 = -112 - 2$; simplify: $2x = -114$. Divide each side of the equation by 2: $\frac{2x}{2} = \frac{-114}{2}$. The variable is now alone: $x = -57$. Therefore the larger value is: $x + 2 = -55$.

292. c. Three consecutive **even** integers are numbers in order like 4, 6, and 8 or –30, –28 and –26, which are each 2 numbers apart. Let x = the first and smallest consecutive even integer. Let $x + 2$ = the second consecutive even integer. Let $x + 4$ = the third and largest consecutive even integer. *Sum* is a key word for addition so the equation becomes $(x) + (x + 2) + (x + 4) = 114$. Combine like terms on the left side of the equation: $3x + 6 = 114$. Subtract 6 from both sides of the equation: $3x + 6 - 6 = 114 - 6$; simplify: $3x = 108$. Divide each side of the equation by 3: $\frac{3x}{3} = \frac{108}{3}$. The variable is now alone: $x = 36$; therefore the next larger integer is: $x + 2 = 38$. The largest even integer would be: $x + 4 = 40$.

293. **d.** Let t = the amount of time traveled. Using the formula *distance = rate ×
time*, substitute the rates of each car and multiply by t to find the distance
traveled by each car. Therefore, $63t$ = distance traveled by one car and
$59t$ = distance traveled by the other car. Since the cars are traveling in
opposite directions, the total distance traveled by both cars is the sum of
these distances: $63t + 59t$. Set this equal to the total distance of 610
miles: $63t + 59t = 610$. Combine like terms on the left side of the
equation: $122t = 610$. Divide each side of the equation by 122: $\frac{122t}{122} = \frac{610}{122}$;
the variable is now alone: $t = 5$. In 5 hours, the cars will be 610 miles
apart.

294. **d.** Use the formula *distance = rate × time* for each train and add these values
together so that the distance equals 822 miles. For the first train, $d = 65t$
and for the second train $d = 72t$, where d is the distance and t is the time
in hours. Add the distances and set them equal to 822: $65t + 72t = 822$.
Combine like terms on the left side of the equation: $137t = 822$; divide
both sides of the equation by 137: $\frac{137t}{137} = \frac{822}{137}$. The variable is now alone:
$t = 6$. In 6 hours, they will be 822 miles apart.

295. **d.** Use the formula *distance = rate × time* for each train and add these values
together so that the distance equals 1,029 miles. For the first train, $d =
45t$ and for the second train $d = 53t$, where d is the distance and t is the
time in hours. Add the distances and set them equal to 1,029: $45t + 53t =
1,029$. Combine like terms on the left side of the equation: $98t = 1,029$;
divide both sides of the equation by 98: $\frac{98t}{98} = \frac{1,029}{98}$. The variable is now
alone: $t = 10.5$ hours. The two trains will pass in 10.5 hours.

296. **c.** Translate the sentence, "Nine minus five times a number is no less than
39," into symbols: $9 - 5x \geq 39$. Subtract 9 from both sides of the
inequality: $9 - 9 - 5x \geq 39 - 9$. Simplify: $-5x \geq 30$; divide both sides of
the inequality by -5. Remember that when dividing or multiplying each
side of an inequality by a negative number, the inequality symbol
changes direction: $\frac{-5x}{-5} \leq \frac{30}{-5}$. The variable is now alone: $x \leq -6$.

297. **a.** This problem is an example of a compound inequality, where there is
more than one inequality in the question. In order to solve it, let x = the
total amount of gumdrops Will has. Set up the compound inequality,
and then solve it as two separate inequalities. Therefore, the second
sentence in the problem can be written as: $2 < x - 2 < 6$. The two

inequalities are: $2 < x - 2$ and $x - 2 < 6$. Add 2 to both sides of both inequalities: $2 + 2 < x - 2 + 2$ and $x - 2 + 2 < 6 + 2$; simplify: $4 < x$ and $x < 8$. If x is greater than four and less than eight, it means that the solution is between 4 and 8. This can be shortened to: $4 < x < 8$.

298. **a.** This inequality shows a solution set where y is greater than or equal to 3 and less than or equal to eight. Both -3 and 8 are in the solution set because of the word *inclusive*, which includes them. The only choice that shows values between -3 and 8 and also includes them is choice **a**.

299. **b.** Let x = the number. Remember that *quotient* is a key word for division, and *at least* means greater than or equal to. From the question, the sentence would translate to: $\frac{x}{2} + 5 \geq x$. Subtract 5 from both sides of the inequality: $\frac{x}{2} + 5 - 5 \geq x - 5$; simplify: $\frac{x}{2} \geq x - 5$. Multiply both sides of the inequality by 2: $\frac{x}{2} \times 2 \geq (x - 5) \times 2$; simplify: $x \geq (x - 5)2$. Use the distributive property on the right side of the inequality: $x \geq 2x - 10$. Add 10 to both sides of the inequality: $x + 10 \geq 2x - 10 + 10$; simplify: $x + 10 \geq 2x$. Subtract x from both sides of the inequality: $x - x + 10 \geq 2x - x$. The variable is now alone: $10 \geq x$. The number is at most 10.

300. **d.** Let x = the amount of hours Cindy worked. Let $2x + 3$ = the amount of hours Carl worked. Since the total hours added together was at most 48, the inequality would be $(x) + (2x + 3) \leq 48$. Combine like terms on the left side of the inequality: $3x + 3 \leq 48$. Subtract 3 from both sides of the inequality: $3x + 3 - 3 \leq 48 - 3$; simplify: $3x \leq 45$. Divide both sides of the inequality by 3: $\frac{3x}{3} \leq \frac{45}{3}$; the variable is now alone: $x \leq 15$. The maximum amount of hours Cindy worked was 15.

301. **b.** Choices **a** and **d** should be omitted because the negative values should not make sense for this problem using time and cost. Choice **b** substituted would be $6 = 2(2) + 2$ which simplifies to $6 = 4 + 2$. Thus, $6 = 6$. The coordinates in choice **c** are reversed from choice **b** and will not work if substituted for x and y.

302. **a.** Let x = the total minutes of the call. Therefore, $x - 1$ = the additional minutes of the call. This choice is correct because in order to calculate the cost, the charge is 35 cents plus 15 cents times the number of additional minutes. If y represents the total cost, then y equals 0.35 plus 0.15 times the quantity $x - 1$. This translates to $y = 0.35 + 0.15(x - 1)$ or $y = 0.15(x - 1) + 0.35$.

303. d. Let x = the total miles of the ride. Therefore, $x - 1$ = the additional miles of the ride. The correct equation takes \$1.25 and adds it to \$1.15 times the number of additional miles, $x - 1$. Translating, this becomes y (the total cost) $= 1.25 + 1.15(x - 1)$, which is the same equation as $y = 1.15(x - 1) + 1.25$.

304. c. The total amount will be \$4.85 plus two times the number of ounces, x. This translates to $4.85 + 2x$, which is the same as $2x + 4.85$. This value needs to be less than or equal to \$10, which can be written as $2x + 4.85 \leq 10$.

305. b. Let x = the number of checks written that month. Green Bank's fees would therefore be represented by $.10x + 3$ and Savings-R-Us would be represented by $.05x + 4.50$. To find the value for which the banks charge the same amount, set the two expressions equal to each other: $.10x + 3 = .05x + 4.50$. Subtract 3 from both sides: $.10x + 3 - 3 = .05x + 4.50 - 3$. This now becomes: $.10x = .05x + 1.50$. Subtract $.05x$ from both sides of the equation: $.10x - .05x = .05x - .05x + 1.50$; this simplifies to: $.05x = 1.50$. Divide both sides of the equation by $.05$: $\frac{.05x}{.05} = \frac{150}{.05}$. The variable is now alone: $x = 30$. Costs would be the same if 30 checks were written.

306. d. Let x = the number of miles traveled in the taxi. The expression for the cost of a ride with Easy Rider would be $1.25x + 2$. The expression for the cost of a ride with Luxury Limo is $1x + 3.25$. To solve, set the two expressions equal to each other: $1.25x + 2 = 1x + 3.25$. Subtract 2 from both sides: $1.25x + 2 - 2 = 1x + 3.25 - 2$. This simplifies to: $1.25x = 1x + 1.25$; subtract $1x$ from both sides: $1.25x - 1x = 1x - 1x + 1.25$. Divide both sides of the equation by $.25$: $\frac{.25x}{.25} = \frac{1.25}{.25}$. The variable is now alone: $x = 5$; the cost would be the same if the trip were 5 miles long.

307. b. Let x = the first integer and let y = the second integer. The equation for the sum of the two integers is $x + y = 36$, and the equation for the difference between the two integers is $x - y = 6$. To solve these by the elimination method, combine like terms vertically and the variable of y cancels out.

$$x + y = 36$$
$$\underline{x - y = 6}$$

This results in: $2x = 42$, so $x = 21$
Substitute the value of x into the first equation to get $21 + y = 36$. Subtract 21 from both sides of this equation to get an answer of $y = 15$.

308. c. Let x = the greater integer and y = the lesser integer. From the first sentence in the question we get the equation $x = y + 2$. From the second sentence in the question we get $y + 2x = 7$. Substitute $x = y + 2$ into the second equation: $y + 2(y + 2) = 7$; use the distributive property to simplify to: $y + 2y + 4 = 7$. Combine like terms to get: $3y + 4 = 7$; subtract 4 from both sides of the equation: $3y + 4 - 4 = 7 - 4$. Simplify to $3y = 3$. Divide both sides of the equation by 3: $\frac{3y}{3} = \frac{3}{3}$; therefore $y = 1$. Since the greater is two more than the lesser, the greater is $1 + 2 = 3$.

309. d. Let x = the lesser integer and let y = the greater integer. The first sentence in the question gives the equation $y = 4x$. The second sentence gives the equation $x + y = 65$. Substitute $y = 4x$ into the second equation: $x + 4x = 65$. Combine like terms on the left side of the equation: $5x = 65$; divide both sides of the equation by 5: $\frac{5x}{5} = \frac{65}{5}$. This gives a solution of $x = 13$, which is the lesser integer.

310. a. Let x = the lesser integer and let y = the greater integer. The first sentence in the question gives the equation $3y + 5x = 9$. The second sentence gives the equation $y - 3 = x$. Substitute $y - 3$ for x in the second equation: $3y + 5(y - 3) = 9$. Use the distributive property on the left side of the equation: $3y + 5y - 15 = 9$. Combine like terms on the left side: $8y - 15 = 9$; add 15 to both sides of the equation: $8y - 15 + 15 = 9 + 15$. Simplify to: $8y = 24$. Divide both sides of the equation by 8: $\frac{8y}{8} = \frac{24}{8}$. This gives a solution of $y = 3$. Therefore the lesser, x, is three less than y, so $x = 0$.

311. b. Let l = the length of the rectangle and let w = the width of the rectangle. Since the width is 6 inches less than 3 times the length, one equation is $w = 3l - 6$. The formula for the perimeter of a rectangle is $2l + 2w = 104$. Substituting the first equation into the perimeter equation for w results in $2l + 2(3l - 6) = 104$. Use the distributive property on the left side of the equation: $2l + 6l - 12 = 104$. Combine like terms on the left side of the equation: $8l - 12 = 104$; add 12 to both sides of the equation: $8l - 12 + 12 = 104 + 12$. Simplify to: $8l = 116$. Divide both sides of the equation by 8: $\frac{8l}{8} = \frac{116}{8}$. Therefore, the length is $l = 14.5$ inches and the width is $w = 3(14.5) - 6 = 37.5$ inches.

312. **a.** Let w = the width of the parallelogram and let l = the length of the parallelogram. Since the length is 5 more than the width, then $l = w + 5$. The formula for the perimeter of a parallelogram $2l + 2w = 50$. Substituting the first equation into the second for l results in $2(w + 5) + 2w = 50$. Use the distributive property on the left side of the equation: $2w + 10 + 2w = 50$; combine like terms on the left side of the equation: $4w + 10 = 50$. Subtract 10 on both sides of the equation: $4w + 10 - 10 = 50 - 10$. Simply to: $4w = 40$. Divide both sides of the equation by 4: $\frac{4w}{4} = \frac{40}{4}$; $w = 10$. Therefore, the width is 10 cm and the length is $10 + 5 = 15$ cm.

313. **c.** Let x = the amount invested at 12% interest. Let y = the amount invested at 15% interest. Since the amount invested at 15% is 100 more then twice the amount at 12%, then $y = 2x + 100$. Since the total interest was $855, use the equation $0.12x + 0.15y = 855$. You have two equations with two variables. Use the second equation $0.12x + 0.15y = 855$ and substitute $(2x + 100)$ for y: $0.12x + 0.15(2x + 100) = 855$. Use the distributive property: $0.12x + 0.3x + 15 = 855$. Combine like terms: $0.42x + 15 = 855$. Subtract 15 from both sides: $0.42x + 15 - 15 = 855 - 15$; simplify: $0.42x = 840$. Divide both sides by 0.42: $\frac{0.42x}{0.42} = \frac{840}{0.42}$. Therefore, $x = \$2,000$, which is the amount invested at 12% interest.

314. **c.** Let x = the amount invested at 8% interest. Since the total interest is $405.50, use the equation $0.06(4,000) + 0.08x = 405.50$. Simplify the multiplication: $240 + 0.08x = 405.50$. Subtract 240 from both sides: $240 - 240 + 0.8x = 405.50 - 240$; simplify: $0.08x = 165.50$. Divide both sides by 0.08: $\frac{0.08x}{0.08} = \frac{165.50}{0.08}$. Therefore, $x = \$2,068.75$, which is the amount invested at 8% interest.

315. **d.** Let x = the amount of coffee at $3 per pound. Let y = the total amount of coffee purchased. If there are 18 pounds of coffee at $2.50 per pound, then the total amount of coffee can be expressed as $y = x + 18$. Use the equation $3x + 2.50(18) = 2.85y$ since the average cost of the y pounds of coffe is $2.85 per pound. To solve, substitute $y = x + 18$ into $3x + 2.50(18) = 2.85y$. $3x + 2.50(18) = 2.85(x + 18)$. Multiply on the left side and use the distributive property on the right side: $3x + 45 = 2.85x + 51.30$. Subtract 2.85x on both sides: $3x - 2.85x + 45 = 2.85x - 2.85x + 51.30$. Simplify: $0.15x + 45 = 51.30$. Subtract 45 from both sides: $0.15x$

$+ 45 - 45 = 51.30 - 45$. Simplify: $0.15x = 6.30$. Divide both sides by 0.15: $\frac{0.15x}{0.15} = \frac{6.30}{0.15}$; so, $x = 42$ pounds, which is the amount of coffee that costs $3 per pound. Therefore, the total amount of coffee is $42 + 18$, which is 60 pounds.

316. c. Let x = the amount of candy at $1.90 per pound. Let y = the total number of pounds of candy purchased. If it is known that there are 40 pounds of candy at $2.15 per pound, then the total amount of candy can be expressed as $y = x + 40$. Use the equation $1.90x + 2.15(40) = \$162$ since the total amount of money spent was $162. Multiply on the left side: $1.90x + 86 = 162$. Subtract 86 from both sides: $1.90x + 86 - 86 = 162 - 86$. Simplify: $1.90x = 76$. Divide both sides by 1.90: $\frac{1.90x}{1.90} = \frac{76}{1.90}$; so, $x = 40$ pounds, which is the amount of candy that costs $1.90 per pound. Therefore, the total amount of candy is $40 + 40$, which is 80 pounds.

317. a. Let x = the amount of marigolds at $1 per packet. Let y = the amount of marigolds at $1.26 per packet. Since there are 50 more packets of the $1.26 seeds than the $1 seeds, $y = x + 50$. Use the equation $1x + 1.26y = 420$ to find the total number of packets of each. By substituting into the second equation, you get $1x + 1.26(x + 50) = 402$. Multiply on the left side using the distributive property: $1x + 1.26x + 63 = 402$. Combine like terms on the left side: $2.26x + 63 = 402$. Subtract 63 from both sides: $2.26x + 63 - 63 = 402 - 63$. Simplify: $2.26x = 339$. Divide both sides by 2.26: $\frac{2.26x}{2.26} = \frac{339}{2.26}$; so, $x = 150$ packets, which is the number of packets that costs $1 each.

318. a. Let x = the amount of 3% iodine solution. Let y = the amount of 20% iodine solution. Since the total amount of solution was 85 oz., then $x + y = 85$. The amount of each type of solution added together and set equal to the amount of 19% solution can be expressed in the equation $0.03x + 0.20y = 0.19(85)$; Use both equations to solve for x. Multiply the second equation by 100 to eliminate the decimal point: $3x + 20y = 19(85)$. Simplify that equation: $3x + 20y = 1805$. Multiply the first equation by -20: $-20x + -20y = -1700$. Add the two equations to eliminate y: $-17x + 0y = -85$. Divide both sides of the equation by -17: $-\frac{17x}{-17} = \frac{-85}{-17}$; $x = 5$. The amount of 3% iodine solution is 5 ounces.

319. **c.** Let x = the amount of 34% acid solution. Let y = the amount of 18% iodine solution. Since the total amount of solution was 30 oz., then $x + y = 30$. The amount of each type of solution added together and set equal to the amount of 28% solution can be expressed in the equation $0.34x + 0.18y = 0.28(30)$. Use both equations to solve for x. Multiply the second equation by 100 to eliminate the decimal point: $34x + 18y = 28(30)$; simplify that equation: $34x + 22y = 840$. Multiply the first equation by -18: $-18x + -18y = -540$. Add the two equations to eliminate y: $16x + 0y = 300$. Divide both sides of the equation by 16: $\frac{16x}{16} = \frac{300}{16}$, $x = 18.75$. The amount of 34% acid solution is 18.75 ounces.

320. **b.** Let x = Ellen's age and let y = Bob's age. Since Bob is 2 years from being twice as old as Ellen, than $y = 2x - 2$. The sum of twice Bob's age and three times Ellen's age is 66 and gives a second equation of $2y + 3x = 66$. Substituting the first equation for y into the second equation results in $2(2x - 2) + 3x = 66$. Use the distributive property on the left side of the equation: $4x - 4 + 3x = 66$; combine like terms on the left side of the equation: $7x - 4 = 66$. Add 4 to both sides of the equation: $7x - 4 + 4 = 66 + 4$. Simplify: $7x = 70$. Divide both sides of the equation by 7: $\frac{7x}{7} = \frac{70}{7}$. The variable, x, is now alone: $x = 10$. Therefore, Ellen is 10 years old.

321. **d.** Let x = Shari's age and let y = Sam's age. Since Sam's age is 1 less than twice Shari's age this gives the equation $y = 2x - 1$. Since the sum of their ages is 104, this gives a second equation of $x + y = 92$. By substituting the first equation into the second for y, this results in the equation $x + 2x - 1 = 92$. Combine like terms on the left side of the equation: $3x - 1 = 92$. Add 1 to both sides of the equation: $3x - 1 + 1 = 92 + 1$. Simplify: $3x = 93$. Divide both sides of the equation by 3: $\frac{3x}{3} = \frac{93}{3}$. The variable, x, is now alone: $x = 31$. Therefore, Shari's age is 31.

322. **d.** Let x = the cost of one binder and let y = the cost of one pen. The first statement, "two binders and three pens cost $12.50," translates to the equation $2x + 3y = 12.50$. The second statement, "three binders and five pens cost $19.50," translates to the equation:

	$3x + 5y = 19.50$
Multiply the first equation by 3:	$6x + 9y = 37.50$
Multiply the second equation by –2:	$-6x + -10y = -39.00$
Combine the two equations to eliminate x:	$-1y = -1.50$
Divide by -1:	$y = 1.50$

Therefore, the cost of one pen is $1.50. Since the cost of 2 binders and 3 pens is 12.50, substitute $y = 1.50$ into the first equation: $3 \times \$1.50 = \$4.50$; $\$12.50 - 4.50 = \$8.00$; $\$8.00 \div 2 = \$4.00$, so each binder is $4.00. The total cost of 1 binder and 1 pen is $4.00 + \$1.50 = \$5.50$.

323. **a.** Let $x =$ the number of degrees in the smaller angle and let $y =$ the number of degrees in the larger angle. Since the angles are complementary, $x + y = 90$. In addition, since the larger angle is 15 more than twice the smaller, $y = 2x + 15$. Substitute the second equation into the first equation for y: $x + 2x + 15 = 90$. Combine like terms on the left side of the equation: $3x + 15 = 90$. Subtract 15 from both sides of the equation: $3x + 15 - 15 = 90 - 15$; simplify: $3x = 75$. Divide both sides by 3: $\frac{3x}{3} = \frac{75}{3}$. The variable, x, is now alone: $x = 25$. The number of degrees in the smaller angle is 25.

324. **b.** Let $x =$ the cost of a student ticket. Let $y =$ the cost of an adult ticket. The first sentence, "The cost of a student ticket is $1 more than half of an adult ticket," gives the equation $x = \frac{1}{2}y + 1$; the second sentence, "six adults and four student tickets cost $76," gives the equation $6y + 4x = 76$. Substitute the first equation into the second for x: $6y + 4(\frac{1}{2}y + 1) = 76$. Use the distributive property on the left side of the equation: $6y + 2y + 4 = 76$. Combine like terms: $8y + 4 = 76$. Subtract 4 on both sides of the equation: $8y + 4 - 4 = 76 - 4$; simplify: $8y = 72$. Divide both sides by 8: $\frac{8y}{8} = \frac{72}{8}$. The variable is now alone: $y = 9$. The cost of one adult ticket is $9.

325. **a.** Let $x =$ the cost of one shirt. Let $y =$ the cost of one tie. The first part of the question, "three shirts and 5 ties cost $23," gives the equation $3x + 5y = 23$; the second part of the question, "5 shirts and one tie cost $20," gives the equation $5x + 1y = 20$. Multiply the second equation by -5: $-25x - 5y = -100$. Add the first equation to that result to eliminate y. The combined equation is: $-22x = -77$. Divide both sides of the equation by -22: $\frac{-22x}{-22} = \frac{-77}{-22}$. The variable is now alone: $x = 3.50$; the cost of one shirt is $3.50.

326. **c.** The terms $3x$ and $5x$ are like terms because they have exactly the same variable with the same exponent. Therefore, you just add the coefficients and keep the variable. $3x + 5x = 8x$.

327. c. Because the question asks for the *difference* between the areas, you need to subtract the expressions: $6a + 2 - 5a$. Subtract like terms: $6a - 5a + 2 = 1a + 2$; $1a = a$, so the simplified answer is $a + 2$.

328. b. Since the area of a rectangle is $A =$ *length times width*, multiply $(x^3)(x^4)$. When multiplying like bases, add the exponents: $x^{3+4} = x^7$.

329. c. Since the area of the soccer field would be found by the formula $A =$ *length* × *width*, multiply the dimensions together: $7y^2 \times 3xy$. Use the commutative property to arrange like variables and the coefficients next to each other: $7 \times 3 \times x \times y^2 \times y$. Multiply: remember that $y^2 \times y = y^2 \times y^1 = y^{2+1} = y^3$. The answer is $21xy^3$.

330. a. Since the area of a parallelogram is $A =$ *base times height*, then the area divided by the base would give you the height; $\frac{x^8}{x^4}$; when dividing like bases, subtract the exponents; $x^{8-4} = x^4$.

331. d. The key word *quotient* means division so the problem becomes $\frac{3d^4}{12d^6}$. Divide the coefficients: $\frac{1d^4}{4d^6}$. When dividing like bases, subtract the exponents: $\frac{1}{4}d^{4-6}$; simplify: $\frac{1d^{-2}}{4}$. A variable in the numerator with a negative exponent is equal to the same variable in the denominator with the opposite sign—in this case, a positive sign on the exponent: $\frac{1}{4d^2}$.

332. a. The translation of the question is $\frac{6x^2 \cdot 4xy^2}{3x^3y}$. The key word *product* tells you to multiply $6x^2$ and $4xy^2$. The result is then divided by $3x^3y$. Use the commutative property in the numerator to arrange like variables and the coefficients together: $\frac{6 \times 4x^2xy^2}{3x^3y}$. Multiply in the numerator. Remember that $x^2 \cdot x = x^2 \cdot x^1 = x^{2+1} = x^3$: $\frac{24x^3y^2}{3x^3y}$. Divide the coefficients; $24 \div 3 = 8$: $\frac{8x^3y^2}{x^3y}$. Divide the variables by subtracting the exponents: $8x^{3-3}y^{2-1}$; simplify. Recall that anything to the zero power is equal to 1: $8x^0y^1 = 8y$.

333. b. Since the formula for the area of a square is $A = s^2$, then by substituting $A = (a^2b^3)^2$. Multiply the outer exponent by each exponent inside the parentheses: $a^{2\times2}b^{3\times2}$. Simplify; a^4b^6.

334. a. The statement in the question would translate to $3x^2(2x^3y)^4$. The word *quantity* reminds you to put that part of the expression in parentheses. Evaluate the exponent by multiplying each number or variable inside the parentheses by the exponent outside the parentheses: $3x^2(2^4x^{3\times4}y^4)$; simplify: $3x^2(16x^{12}y^4)$. Multiply the coefficients and add the exponents of like variables: $3(16x^{2+12}y^4)$; simplify: $48x^{14}y^4$.

335. d. Since the area of a rectangle is $A = length\ times\ width$, multiply the dimensions to find the area: $2x(4x + 5)$. Use the distributive property to multiply each term inside the parentheses by $2x$: $2x \times 4x + 2x \times 5$. Simplify by multiplying the coefficients of each term and adding the exponents of the like variables: $8x^2 + 10x$.

336. b. The translated expression would be $-9p^3r(5p - 6r)$. Remember that the key word *product* means multiply. Use the distributive property to multiply each term inside the parentheses by $-9p^3r$: $-9p^3r \times 5p - (-9p^3r) \times 6r$. Simplify by multiplying the coefficients of each term and adding the exponents of the like variables: $-9 \times 5p^{3+1}r - (-9 \times 6p^3r^{1+1})$. Simplify: $-45p^4r - (-54p^3r^2)$. Change subtraction to addition and change the sign of the following term to its opposite: $-45p^4r + (+54p^3r^2)$; this simplifies to: $-45p^4r + 54p^3r^2$.

337. c. The two numbers in terms of x would be $x + 3$ and $x + 4$ since *increased by* would tell you to add. *Product* tells you to multiply these two quantities: $(x + 3)(x + 4)$. Use **FOIL** (**F**irst terms of each binomial multiplied, **O**uter terms in each multiplied, **I**nner terms of each multiplied, and **L**ast term of each binomial multiplied) to multiply the binomials: $(x \cdot x) + (4 \cdot x) + (3 \cdot x) + (3 \cdot 4)$; simplify each term: $x^2 + 4x + 3x + 12$. Combine like terms: $x^2 + 7x + 12$.

338. d. Since the area of a rectangle is $A = length\ times\ width$, multiply the two expressions together: $(2x - 1)(x + 6)$. Use **FOIL** (**F**irst terms of each binomial multiplied, **O**uter terms in each multiplied, **I**nner terms of each multiplied, and **L**ast term of each binomial multiplied) to multiply the binomials: $(2x \cdot x) + (2x \cdot 6) - (1 \cdot x) - (1 \cdot 6)$. Simplify: $2x^2 + 12x - x - 6$; combine like terms: $2x^2 + 11x - 6$.

339. a. Use the formula *distance = rate × time*. By substitution, distance $= (4x^2 - 2) \times (3x - 8)$. Use **FOIL** (**F**irst terms of each binomial multiplied, **O**uter terms in each multiplied, **I**nner terms of each multiplied, and **L**ast term of each binomial multiplied) to multiply the binomials: $(4x^2 \cdot 3x) - (8 \cdot 4x^2) - (2 \cdot 3x) - (2 \cdot -8)$. Simplify each term: $12x^3 - 32x^2 - 6x + 16$.

340. d. Since the formula for the volume of a prism is $V = Bh$, where B is the area of the base and h is the height of the prism, $V = (x - 3)(x^2 + 4x + 1)$. Use the distributive property to multiply the first term of the binomial, x, by each term of the trinomial, and then the second term of the binomial, -3, by each term of the trinomial: $x(x^2 + 4x + 1) - 3(x^2 + 4x + 1)$. Then distribute: $(x \cdot x^2) + (x \cdot 4x) + (x \cdot 1) - (3 \cdot x^2) - (3 \cdot 4x) - (3 \cdot 1)$. Simplify by multiplying within each term: $x^3 + 4x^2 + x - 3x^2 - 12x - 3$. Use the commutative property to arrange like terms next to each other. Remember that $1x = x$: $x^3 + 4x^2 - 3x^2 + x - 12x - 3$; combine like terms: $x^3 + x^2 - 11x - 3$.

341. b. Since the formula for the volume of a rectangular prism is $V = l \times w \times h$, multiply the dimensions together: $(x + 1)(x - 2)(x + 4)$. Use **FOIL** (**F**irst terms of each binomial multiplied, **O**uter terms in each multiplied, **I**nner terms of each multiplied, and **L**ast term of each binomial multiplied) to multiply the first two binomials: $(x + 1)(x - 2)$; $(x \cdot x) + x(-2) + (1 \cdot x) + 1(-2)$. Simplify by multiplying within each term: $x^2 - 2x + 1x - 2$; combine like terms: $x^2 - x - 2$. Multiply the third factor by this result: $(x + 4)(x^2 - x - 2)$. To do this, use the distributive property to multiply the first term of the binomial, x, by each term of the trinomial, and then the second term of the binomial, 4, by each term of the trinomial: $x(x^2 - x - 2) + 4(x^2 - x - 2)$. Distribute: $(x \cdot x^2) + (x \cdot -x) + (x \cdot -2) + (4 \cdot x^2) + (4 \cdot -x) + (4 \cdot -2)$. Simplify by multiplying in each term: $x^3 - x^2 - 2x + 4x^2 - 4x - 8$. Use the commutative property to arrange like terms next to each other: $x^3 - x^2 + 4x^2 - 2x - 4x - 8$; combine like terms: $x^3 + 3x^2 - 6x - 8$.

342. c. Since area of a rectangle is found by multiplying length by width, we need to find the factors that multiply out to yield $x^2 - 36$. Because x^2 and 36 are both perfect squares ($x^2 = x \cdot x$ and $36 = 6 \cdot 6$), the product, $x^2 - 36$, is called a difference of two perfect squares, and its factors are the sum and difference of the square roots of its terms. Therefore, because the square root of $x^2 = x$ and the square root of $36 = 6$, $x^2 - 36 = (x + 6)(x - 6)$.

343. b. To find the base and the height of the parallelogram, find the factors of this binomial. First look for factors that both terms have in common; $2x^2$ and $10x$ both have a factor of 2 and x. Factor out the greatest common factor, $2x$, from each term. $2x^2 - 10x$; $2x(x - 5)$. To check an answer like this, multiply through using the distributive property. $2x(x-5)$; $(2x \cdot x) - (2x \cdot 5)$; simplify and look for a result that is the same as the original question. This question checked: $2x^2 - 10x$.

344. **d.** Since the formula for the area of a rectangle is $A = length \ times \ width$, find the two factors of $x^2 + 2x + 1$ to get the dimensions. First check to see if there is a common factor in each of the terms or if it is the difference between two perfect squares, and it is neither of these. The next step would be to factor the trinomial into two binomials. To do this, you will be doing a method that resembles **FOIL** backwards (**F**irst terms of each binomial multiplied, **O**uter terms in each multiplied, **I**nner terms of each multiplied, and **L**ast term of each binomial multiplied.) **F**irst results in x^2, so the first terms must be: $(x \ )(x \ )$; **O**uter added to the **I**nner combines to $2x$, and the **L**ast is 1, so you need to find two numbers that add to +2 and multiply to +1. These two numbers would have to be +1 and +1: $(x + 1)(x + 1)$. Since the factors of the trinomial are the same, this is an example of a perfect square trinomial, meaning that the farmer's rectangular field was, more specifically, a square field. To check to make sure these are the factors, multiply them by using **FOIL** (**F**irst terms of each binomial multiplied, **O**uter terms in each multiplied, **I**nner terms of each multiplied, and **L**ast term of each binomial multiplied; $(x \cdot x) + (1 \cdot x) + (1 \cdot x) + (1 \cdot 1)$; multiply in each term: $x^2 + 1x + 1x + 1$; combine like terms: $x^2 + 2x + 1$.

345. **a.** Since area of a rectangle is $length \times width$, look for the factors of the trinomial to find the two dimensions. First check to see if there is a common factor in each of the terms or if it is the difference between two perfect squares, and it is neither of these. The next step would be to factor the trinomial into two binomials. To do this, you will be doing a method that resembles **FOIL** backwards. (**F**irst terms of each binomial multiplied, **O**uter terms in each multiplied, **I**nner terms of each multiplied, and **L**ast term of each binomial multiplied.) **F**irst results in x^2, so the first terms must be $(x \ )(x \ )$; **O**uter added to the **I**nner combines to $6x$, and the **L**ast is 5, so you need to find two numbers that add to produce +6 and multiply to produce +5. These two numbers are +1 and +5; $(x + 1)(x + 5)$.

346. **a.** Since the formula for the area of a rectangle is *length* × *width*, find the factors of the trinomial to get the dimensions. First check to see if there is a common factor in each of the terms or if it is the difference between two perfect squares, and it is neither of these. The next step would be to factor the trinomial into two binomials. To do this, you will be doing a method that resembles **FOIL** backwards (**F**irst terms of each binomial multiplied, **O**uter terms in each multiplied, **I**nner terms of each multiplied, and **L**ast term of each binomial multiplied.) **F**irst results in x^2, so the first terms must be $(x\ )(x\ )$; **O**uter added to the **I**nner combines to $1x$, and the **L**ast is -12, so you need to find two numbers that add to $+1$ and multiply to -12. These two numbers are -3 and $+4$; $(x - 3)(x + 4)$. Thus, the dimensions are $(x + 4)$ and $(x - 3)$.

347. **b.** Since the trinomial does not have a coefficient of one on its highest exponent term, the easiest way to find the answer to this problem is to use the distributive property. First, using the original trinomial, identify the sum and product by looking at the terms when the trinomial is in descending order (highest exponent first): $3x^2 - 7x + 2$. The sum is the middle term, in this case, $-7x$. The product is the product of the first and last terms, in this case, $(3x^2)(2) = 6x^2$. Now, identify two quantities whose sum is $-7x$ and product is $6x^2$, namely $-6x$ and $-x$. Rewrite the original trinomial using these two terms to replace the middle term in any order: $3x^2 - 6x - x + 2$. Now factor by grouping by taking a common factor out of each pair of terms. The common factor of $3x^2$ and $-6x$ is $3x$ and the common factor of $-x$ and 2 is -1. Therefore, $3x^2 - 6x - x + 2$ becomes $3x(x - 2) - 1(x - 2)$. (Notice that if this expression were multiplied back out and simplified, it would correctly yield the original polynomial.) Now, this two-term expression has a common factor of $(x - 2)$, which can be factored out of each term using the distributive property: $3x(x - 2) - 1(x - 2)$ becomes $(x - 2)(3x - 1)$. The dimensions of the courtyard are $(x - 2)$ and $(3x - 1)$.

348. **d.** In order to convert this number to standard notation, multiply 9.3 by the factor of 10^7. Since 10^7 is equal to 10,000,000, $9.3 \times 10,000,000$ is equal to 93,000,000. As an equivalent solution, move the decimal point in 9.3 seven places to the right since the exponent on the 10 is positive 7.

349. **c.** To convert to scientific notation, place a decimal point after the first non-zero digit to create a number between 1 and 10—in this case, between the 2 and the 4. Count the number of decimal places from that decimal to the place of the decimal in the original number. In this case, the number of places would be 5. This number, 5, becomes the exponent of 10 and is positive because the original number was greater than one. The answer then is 2.4×10^5.

350. **d.** In order to convert this number to standard notation, multiply 5.3 by the factor of 10^{-6}. Since 10^{-6} is equal to 0.000001, 5.3×0.000001 is equal to 0.0000053. Equivalently, move the decimal point in 5.3 six places to the left since the exponent on the 10 is negative 6.

351. **d.** Let x = the number. The sentence, "The square of a positive number is 49," translates to the equation $x^2 = 49$. Take the square root of each side to get $\sqrt{x^2} = \sqrt{49}$ so $x = 7$ or -7. Since you are looking for a positive number, the final solution is 7.

352. **d.** Let x = the number. The statement, "The square of a number added to 25 equals 10 times the number," translates to the equation $x^2 + 25 = 10x$. Put the equation in standard form $ax^2 + bx + c = 0$, and set it equal to zero: $x^2 - 10x + 25 = 0$. Factor the left side of the equation: $(x - 5)(x - 5) = 0$. Set each factor equal to zero and solve: $x - 5 = 0$ or $x - 5 = 0$; $x = 5$ or $x = 5$. The number is 5.

353. **b.** Let x = the number. The statement, "The sum of the square of a number and 12 times the number is -27," translates to the equation $x^2 + 12x = -27$. Put the equation in standard form and set it equal to zero: $x^2 + 12x + 27 = 0$. Factor the left side of the equation: $(x + 3)(x + 9) = 0$. Set each factor equal to zero and solve: $x + 3 = 0$ or $x + 9 = 0$; $x = -3$ or $x = -9$. The possible values of this number are -3 or -9, the smaller of which is -9.

354. **b.** Let x = the number of inches in the width and let $x + 5$ = the number of inches in the length. Since area of a rectangle is *length times width*, the equation for the area of the rectangle is $x(x + 5) = 24$. Multiply the left side of the equation using the distributive property: $x^2 + 5x = 24$. Put the equation in standard form and set it equal to zero: $x^2 + 5x - 24 = 0$. Factor the left side of the equation: $(x + 8)(x - 3) = 0$. Set each factor equal to zero and solve: $x + 8 = 0$ or $x - 3 = 0$; $x = -8$ or $x = 3$. Reject the solution of -8 because a distance will not be negative. The width is 3 inches.

355. b. Let x = the measure of the base and let $x + 5$ = the measure of the height. Since the area of a parallelogram is *base times height*, then the equation for the area of the parallelogram is $x(x + 5) = 36$. Multiply the left side of the equation using the distributive property: $x^2 + 5x = 36$; Put the equation in standard form and set it equal to zero: $x^2 + 5x - 36 = 0$. Factor the left side of the equation: $(x + 9)(x - 4) = 0$. Set each factor equal to zero and solve: $x + 9 = 0$ or $x - 4 = 0$; $x = -9$ or $x = 4$. Reject the solution of -9 because a distance will not be negative. The height is $4 + 5 = 9$ meters.

356. d. Let x = the length of the diagonal. Therefore, $x - 5$ = the length of the patio and $x - 7$ = the width of the patio. Since the area is 195 m^2, and area is *length times the width*, the equation is $(x - 5)(x - 7) = 195$. Use the distributive property to multiply the binomials: $x^2 - 5x - 7x + 35 = 195$. Combine like terms: $x^2 - 12x + 35 = 195$. Subtract 195 from both sides: $x^2 - 12x + 35 - 195 = 195 - 195$. Simplify: $x^2 - 12x - 160 = 0$. Factor the result: $(x - 20)(x + 8) = 0$. Set each factor equal to 0 and solve: $x - 20 = 0$ or $x + 8 = 0$; $x = 20$ or $x = -8$. Reject the solution of -8 because a distance will not be negative. The length of the diagonal is 20 m.

357. a. Let w = the width of the field and let $2w + 2$ = the length of the field (two more than twice the width). Since area is *length times width*, multiply the two expressions together and set them equal to 3,280: $w(2w + 2) = 3,280$. Multiply using the distributive property: $2w^2 + 2w = 3,280$. Subtract 3,280 from both sides: $2w^2 + 2w - 3,280 = 3,280 - 3,280$; simplify: $2w^2 + 2w - 3,280 = 0$. Factor the trinomial completely: $2(w^2 + w - 1640) = 0$; $2(w + 41)(w - 40) = 0$. Set each factor equal to zero and solve: $2 \neq 0$ or $w + 41 = 0$ or $w - 40 = 0$; $w = -41$ or $w = 40$. Reject the negative solution because you will not have a negative width. The width is 40 feet.

358. b. Let x = the width of the walkway. Since the width of the garden only is 24, the width of the garden and the walkway together is $x + x + 24$ or $2x + 24$. Since the length of the garden only is 35, the length of the garden and the walkway together is $x + x + 35$ or $2x + 35$. Area of a rectangle is *length times width*, so multiply the expressions together and set the result equal to the total area of 1,530 square feet: $(2x + 24)(2x + 35) = 1,530$. Multiply the binomials using the distributive property: $4x^2 + 70x + 48x + 840 = 1,530$. Combine like terms: $4x^2 + 118x + 840 = 1,530$. Subtract

1,530 from both sides: $4x^2 + 118x + 840 - 1,530 = 1,530 - 1,530$; simplify: $4x^2 + 118x - 690 = 0$. Factor the trinomial completely: $2(2x^2 + 59x - 345) = 0$; $2(2x + 69)(x - 5) = 0$. Set each factor equal to zero and solve: $2 \neq 0$ or $2x + 69 = 0$ or $x - 5 = 0$; $x = -34.5$ or $x = 5$. Reject the negative solution because you will not have a negative width. The width is 5 feet.

359. a. Let x = the width of the deck. Since the width of the pool only is 18, the width of the pool and the deck is $x + x + 18$ or $2x + 18$. Since the length of the pool only is 24, the length of the pool and the deck together is $x + x + 24$ or $2x + 24$. The total area for the pool and the deck together is 832 square feet, 400 square feet added to 432 square feet for the pool. Area of a rectangle is *length times width* so multiply the expressions together and set them equal to the total area of 832 square feet: $(2x + 18)(2x + 24) = 832$. Multiply the binomials using the distributive property: $4x^2 + 36x + 48x + 432 = 832$. Combine like terms: $4x^2 + 84x + 432 = 832$. Subtract 832 from both sides: $4x^2 + 84x + 432 - 832 = 832 - 832$; simplify: $4x^2 + 84x - 400 = 0$. Factor the trinomial completely: $2(2x^2 + 42x - 200) = 0$; $2(2x - 8)(x + 25) = 0$. Set each factor equal to zero and solve: $2 \neq 0$ or $2x - 8 = 0$ or $x + 25 = 0$; $x = 4$ or $x = -25$. Reject the negative solution because you will not have a negative width. The width is 4 feet.

360. c. To solve this problem, find the width of the frame first. Let x = the width of the frame. Since the width of the picture only is 12, the width of the frame and the picture is $x + x + 12$ or $2x + 12$. Since the length of the picture only is 14, the length of the frame and the picture together is $x + x + 14$ or $2x + 14$. The total area for the frame and the picture together is 288 square inches. Area of a rectangle is *length times width* so multiply the expressions together and set them equal to the total area of 288 square inches: $(2x + 12)(2x + 14) = 288$. Multiply the binomials using the distributive property: $4x^2 + 28x + 24x + 168 = 288$. Combine like terms: $4x^2 + 52x + 168 = 288$. Subtract 288 from both sides: $4x^2 + 52x + 168 - 288 = 288 - 288$; simplify: $4x^2 + 52x - 120 = 0$. Factor the trinomial completely: $4(x^2 + 13x - 30) = 0$; $4(x - 2)(x + 15) = 0$. Set each factor equal to zero and solve: $4 \neq 0$ or $x - 2 = 0$ or $x + 15 = 0$; $x = 2$ or $x = -15$. Reject the negative solution because you will not have a negative width. The width is 2 feet. Therefore, the larger dimension of the frame is $2(2) + 14 = 4 + 14 = 18$ inches.

361. b. Let x = the lesser integer and let $x + 1$ = the greater integer. Since *product* is a key word for multiplication, the equation is $x(x + 1) = 90$. Multiply using the distributive property on the left side of the equation: $x^2 + x = 90$. Put the equation in standard form and set it equal to zero: $x^2 + x - 90 = 0$. Factor the trinomial: $(x - 9)(x + 10) = 0$. Set each factor equal to zero and solve: $x - 9 = 0$ or $x + 10 = 0$; $x = 9$ or $x = -10$. Since you are looking for a positive integer, reject the x-value of -10. Therefore, the lesser positive integer would be 9.

362. a. Let x = the lesser integer and let $x + 1$ = the greater integer. Since *product* is a key word for multiplication, the equation is $x(x + 1) = 210$. Multiply using the distributive property on the left side of the equation: $x^2 + x = 210$. Put the equation in standard form and set it equal to zero: $x^2 + x - 210 = 0$. Factor the trinomial: $(x - 14)(x + 15) = 0$. Set each factor equal to zero and solve: $x - 14 = 0$ or $x + 15 = 0$; $x = 14$ or $x = -15$. Since you are looking for a negative integer, reject the x-value of 14. Therefore, $x = -15$ and $x + 1 = -14$. The greater negative integer is -14.

363. a. Let x = the lesser even integer and let $x + 2$ = the greater even integer. Since *product* is a key word for multiplication, the equation is $x(x + 2) = 168$. Multiply using the distributive property on the left side of the equation: $x^2 + 2x = 168$. Put the equation in standard form and set it equal to zero: $x^2 + 2x - 168 = 0$. Factor the trinomial: $(x - 12)(x + 14) = 0$. Set each factor equal to zero and solve: $x - 12 = 0$ or $x + 14 = 0$; $x = 12$ or $x = -14$. Since you are looking for a positive integer, reject the x-value of -14. Therefore, the lesser positive integer would be 12.

364. d. Let x = the lesser odd integer and let $x + 2$ = the greater odd integer. Since *product* is a key word for multiplication, the equation is $x(x + 2) = 143$. Multiply using the distributive property on the left side of the equation: $x^2 + 2x = 143$. Put the equation in standard form and set it equal to zero: $x^2 + 2x - 143 = 0$. Factor the trinomial: $(x - 11)(x + 13) =$

0. Set each factor equal to zero and solve: $x - 11 = 0$ or $x + 13 = 0$; $x = 11$ or $x = -13$. Since you are looking for a positive integer, reject the x-value of -13. Therefore, $x = 11$ and $x + 2 = 13$. The greater positive odd integer is 13.

365. c. Let x = the lesser odd integer and let $x + 2$ = the greater odd integer. The translation of the sentence, "The sum of the squares of two consecutive odd integers is 74," is the equation $x^2 + (x + 2)^2 = 74$. Multiply $(x + 2)^2$ out as $(x + 2)(x + 2)$ using the distributive property: $x^2 + (x^2 + 2x + 2x + 4) = 74$. Combine like terms on the left side of the equation: $2x^2 + 4x + 4 = 74$. Put the equation in standard form by subtracting 74 from both sides, and set it equal to zero: $2x^2 + 4x - 70 = 0$; factor the trinomial completely: $2(x^2 + 2x - 35) = 0$; $2(x - 5)(x + 7) = 0$. Set each factor equal to zero and solve: $2 \neq 0$ or $x - 5 = 0$ or $x + 7 = 0$; $x = 5$ or $x = -7$. Since you are looking for a positive integer, reject the solution of $x = -7$. Therefore, the smaller positive integer is 5.

366. a. Let x = the lesser integer and let $x + 1$ = the greater integer. The sentence, "the difference between the squares of two consecutive integers is 15," can translate to the equation $(x + 1)^2 - x^2 = 15$. Multiply the binomial $(x + 1)^2$ as $(x + 1)(x + 1)$ using the distributive property: $x^2 + 1x + 1x + 1 - x^2 = 15$. Combine like terms: $2x + 1 = 15$; subtract 1 from both sides of the equation: $2x + 1 - 1 = 15 - 1$. Divide both sides by 2: $\frac{2x}{2} = \frac{14}{2}$. The variable is now alone: $x = 7$. Therefore, the larger consecutive integer is $x + 1 = 8$.

367. c. Let x = the lesser integer and let $x + 1$ = the greater integer. The sentence, "The square of one integer is 55 less than the square of the next consecutive integer," can translate to the equation $x^2 = (x + 1)^2 - 55$. Multiply the binomial $(x + 1)^2$ as $(x + 1)(x + 1)$ using the distributive property: $x^2 = x^2 + 1x + 1x + 1 - 55$. Combine like terms: $x^2 = x^2 + 2x - 54$. Subtract x^2 from both sides of the equation: $x^2 - x^2 = x^2 - x^2 + 2x - 54$. Add 54 to both sides of the equation: $0 + 54 = 2x - 54 + 54$. Divide both sides by 2: $\frac{54}{2} = \frac{2x}{2}$. The variable is now alone: $27 = x$. The lesser integer is 27.

368. c. Let x = the amount each side is increased. Then, $x + 4$ = the new width and $x + 6$ = the new length. Since area is *length times width*, the formula using the new area is $(x + 4)(x + 6) = 168$. Multiply using the distributive property on the left side of the equation: $x^2 + 6x + 4x + 24 = 168$; combine like terms: $x^2 + 10x + 24 = 168$. Subtract 168 from both sides: $x^2 + 10x + 24 - 168 = 168 - 168$. Simplify: $x^2 + 10x - 144 = 0$. Factor the trinomial: $(x - 8)(x + 18) = 0$. Set each factor equal to zero and solve: $x - 8 = 0$ or $x + 18 = 0$; $x = 8$ or $x = -18$. Reject the negative solution because you won't have a negative dimension. The correct solution is 8 inches.

369. a. Let x = the amount of reduction. Then $4 - x$ = the width of the reduced picture and $6 - x$ = the length of the reduced picture. Since area is *length times width*, and one-third of the old area of 24 is 8, the equation for the area of the reduced picture would be $(4 - x)(6 - x) = 8$. Multiply the binomials using the distributive property: $24 - 4x - 6x + x^2 = 8$; combine like terms: $24 - 10x + x^2 = 8$. Subtract 8 from both sides: $24 - 8 - 10x + x^2 = 8 - 8$. Simplify and place in standard form: $x^2 - 10x + 16 = 0$. Factor the trinomial into 2 binomials: $(x - 2)(x - 8) = 0$. Set each factor equal to zero and solve: $x - 2 = 0$ or $x - 8 = 0$; $x = 2$ or $x = 8$. The solution of 8 is not reasonable because it is greater than the original dimensions of the picture. Accept the solution of $x = 2$ and the smaller dimension of the reduced picture would be $4 - 2 = 2$ inches.

370. d. Let x = the amount that each side of the garden is increased. Then, $x + 20$ = the new width and $x + 24$ = the new length. Since the area of a rectangle is *length times width*, then the area of the old garden is $20 \times 24 = 480$ and the new area is $480 + 141 = 621$. The equation using the new area becomes $(x + 20)(x + 24) = 621$. Multiply using the distributive property on the left side of the equation: $x^2 + 24x + 20x + 480 = 621$; combine like terms: $x^2 + 44x + 480 = 621$. Subtract 621 from both sides: $x^2 + 44x + 480 - 621 = 621 - 621$; simplify: $x^2 + 44x - 141 = 0$. Factor the trinomial: $(x - 3)(x + 47) = 0$. Set each factor equal to zero and solve: $x - 3 = 0$ or $x + 47 = 0$; $x = 3$ or $x = -47$. Reject the negative solution because you won't have a negative increase. Thus, each side will be increased by 3 and the new length would be $24 + 3 = 27$ feet.

371. **b.** Let x = the number of hours it takes Ian and Jack to remodel the kitchen if they are working together. Since it takes Ian 20 hours if working alone, he will complete $\frac{1}{20}$ of the job in one hour, even when he's working with Jack. Similarly, since it takes Jack 15 hours to remodel a kitchen, he will complete $\frac{1}{15}$ of the job in one hour, even when he's working with Ian. Since it takes x hours for Ian and Jack to complete the job together, it stands to reason that at the end of one hour, their combined effort will have completed $\frac{1}{x}$ of the job. Therefore, *Ian's work + Jack's work = combined work* and we have the equation: $\frac{1}{20} + \frac{1}{15} = \frac{1}{x}$. Multiply through by the least common denominator of 20, 15 and x which is $60x$: $(60x)(\frac{1}{20})$ + $(60x)(\frac{1}{15}) = (60x)(\frac{1}{x})$. Simplify: $3x + 4x = 60$. Simplify: $7x = 60$. Divide by 7: $\frac{7x}{7} = \frac{60}{7}$. $x = \frac{60}{7}$ which is about 8.6 hours.

372. **a.** Let x = the number of hours it takes Peter and Joe to paint a room if they are working together. Since it takes Peter 1.5 hours if working alone, he will complete $\frac{1}{1.5}$ of the job in one hour, even when he's working with Joe. Similarly, since it takes Joe 2 hours to paint a room working alone, he will complete $\frac{1}{2}$ of the job in one hour, even when working with Peter. Since it takes x hours for Peter and Joe to complete the job together, it stands to reason that at the end of one hour, their combined effort will have completed $\frac{1}{x}$ of the job. Therefore, *Peter's work + Joe's work = combined work* and we have the equation: $\frac{1}{1.5} + \frac{1}{2} = \frac{1}{x}$. Multiply through by the least common denominator of 1.5, 2 and x which is $6x$: $(6x)(\frac{1}{1.5}) + (6x)(\frac{1}{2}) = (6x)(\frac{1}{x})$. Simplify: $4x + 3x = 6$. Simplify: $7x = 6$. Divide by 7: $\frac{7x}{7} = \frac{6}{7}$; $x = \frac{6}{7}$ hours. Change hours into minutes by multiplying by 60 since there are 60 minutes in one hour. $(60)(\frac{6}{7}) = \frac{360}{7}$ divided by 7 equals 51.42 minutes which rounds to 51 minutes.

373. c. Let x = the number of hours it takes Carla and Charles to plant a garden if they are working together. Since it takes Carla 3 hours if working alone, she will complete $\frac{1}{3}$ of the job in one hour, even when she's working with Charles. Similarly, since it takes Charles 4.5 hours to plant a garden working alone, he will complete $\frac{1}{4.5}$ of the job in one hour, even when working with Carla. Since it takes x hours for Carla and Charles to complete the job together, it stands to reason that at the end of one hour, their combined effort will have completed $\frac{1}{x}$ of the job. Therefore, *Carla's work + Charles's work = combined work* and we have the equation: $\frac{1}{3} + \frac{1}{4.5} = \frac{1}{x}$. Multiply through by the least common denominator of 3, 4.5 and x which is $9x$: $(9x)(\frac{1}{3}) + (9x)(\frac{1}{4.5}) = (9x)(\frac{1}{x})$. Simplify: $3x + 2x = 9$. Simplify: $5x = 9$. Divide by 5: $\frac{5x}{5} = \frac{9}{5}$; $x = \frac{9}{5}$ hours which is equal to 1.8 hours.

374. c. Let x = the number of hours it will take Jerry to do the job alone. In 1 hour Jim can do $\frac{1}{10}$ of the work, and Jerry can do $\frac{1}{x}$ of the work. As an equation this looks like $\frac{1}{10} + \frac{1}{x} = \frac{1}{4}$, where $\frac{1}{4}$ represents what part of the job they can complete in one hour together. Multiplying both sides of the equation by the least common denominator, $40x$, results in the equation $4x + 40 = 10x$. Subtract $4x$ from both sides of the equation. $4x - 4x + 40 = 10x - 4x$. This simplifies to $40 = 6x$. Divide each side of the equation by 6; $\frac{40}{6} = \frac{6x}{6}$. Therefore, $6.666 = x$, and it would take Jerry about 6.7 hours to complete the job alone.

375. d. Let x = the number of hours Ben takes to clean the garage by himself. In 1 hour Ben can do $\frac{1}{x}$ of the work and Bill can do $\frac{1}{15}$ of the work. As an equation this looks like $\frac{1}{x} + \frac{1}{15} = \frac{1}{6}$, where $\frac{1}{6}$ represents what part they can clean in one hour together. Multiply both sides of the equation by the least common denominator, $30x$, to get an equation of $30 + 2x = 5x$. Subtract $2x$ from both sides of the equation; $30 + 2x - 2x = 5x - 2x$. This simplifies to $30 = 3x$, and dividing both sides by 3 results in a solution of 10 hours.

Geometry

The geometry problems in this chapter involve lines, angles, triangles, rectangles, squares, and circles. You will learn how to find length, perimeter, area, circumference, and volume, and how you can apply geometry to everyday problems.

376. Charlie wants to know the area of his property, which measures 120 ft by 150 ft. Which formula will he use?

    a. $A = s^2$
    b. $A = \pi r^2$
    c. $A = \frac{1}{2}bh$
    d. $A = lw$

377. Dawn wants to compare the volume of a basketball with the volume of a tennis ball. Which formula will she use?

    a. $V = \pi r^2 h$
    b. $V = \frac{4}{3}\pi r^3$
    c. $V = \frac{1}{3}\pi r^2 h$
    d. $V = s^3$

378. Rick is ordering a new triangular sail for his boat. He needs to know the area of the sail. Which formula will he use?

a. $A = lw$

b. $A = \frac{1}{2}bh$

c. $A = bh$

d. $A = \frac{1}{2}h(b_1 + b_2)$

379. Keith wants to know the surface area of a basketball. Which formula will he use?

a. $s = 6s^2$

b. $s = 4\pi r^2$

c. $s = 2\pi r^2 + 2\pi rh$

d. $s = \pi r^2 + 2\pi rh$

380. Aaron is installing a ceiling fan in his bedroom. Once the fan is in motion, he needs to know the area the fan will cover. Which formula will he use?

a. $A = bh$

b. $A = s^2$

c. $A = \frac{1}{2}bh$

d. $A = \pi r^2$

381. Mimi is filling a tennis ball can with water. She wants to know the volume of the cylinder shaped can. What formula will she use?

a. $V = \pi r^2 h$

b. $V = \frac{4}{3}\pi r^3$

c. $V = \frac{1}{3}\pi r^2 h$

d. $V = s^3$

382. Audrey is creating a raised flowerbed that is 4.5 ft by 4.5 ft. She needs to calculate how much lumber to buy. If she needs to know the distance around the flowerbed, which formula is easiest to use?

a. $P = a + b + c$

b. $A = lw$

c. $P = 4s$

d. $C = 2\pi r$

383. Al is painting a right cylinder storage tank. In order to purchase the correct amount of paint he needs to know the total surface area to be painted. Which formula will he use if he does not paint the bottom of the tank?

a. $S = 2\pi r^2 + 2\pi rh$

b. $S = 4\pi r^2$

c. $S = \pi r^2 + 2\pi rh$

d. $S = 6s^2$

384. Cathy is creating a quilt out of fabric panels that are 6 in by 6 in. She wants to know the total area of her square-shaped quilt. Which formula will she use?

a. $A = s^2$

b. $A = \frac{1}{2}bh$

c. $A = \pi r^2$

d. $A = \frac{1}{2}h(b_1 + b_2)$

385. If Lisa wants to know the distance around her circular table, which has a diameter of 42 in, which formula will she use?

a. $P = 4s$

b. $P = 2l + 2w$

c. $C = \pi d$

d. $P = a + b + c$

386. Danielle needs to know the distance around a basketball court. Which geometry formula will she use?

a. $P = 2l + 2w$

b. $P = 4s$

c. $P = a + b + c$

d. $P = b_1 + b_2 + h$

387. To find the volume of a cube that measures 3 cm by 3 cm by 3 cm, which formula would you use?

a. $V = \pi r^2 h$

b. $V = \frac{4}{3}\pi r^3$

c. $V = \frac{1}{3}\pi r^2 h$

d. $V = s^3$

388. To find the perimeter of a triangular region, which formula would you use?

a. $P = a + b + c$

b. $P = 4s$

c. $P = 2l + 2w$

d. $C = 2\pi r$

389. A racquetball court is 40 ft by 20 ft. What is the area of the court in square feet?

a. 60 ft^2

b. 80 ft^2

c. 800 ft^2

d. 120 ft^2

390. Allan has been hired to mow the school soccer field, which is 180 ft wide by 330 ft long. If his mower mows strips that are 2 feet wide, how many times must he mow across the width of the lawn?

a. 90

b. 165

c. 255

d. 60

391. Erin is painting a bathroom with four walls each measuring 9 ft by 5.5 ft. Ignoring the doors or windows, what is the area to be painted?

a. 198 ft^2

b. 66 ft^2

c. 49.5 ft^2

d. 160 ft^2

392. The arm of a ceiling fan measures a length of 25 in. What is the area covered by the motion of the fan blades when turned on? ($\pi = 3.14$)

a. 246.49 in^2

b. 78.5 in^2

c. 1,962.5 in^2

d. 157 in^2

393. A building that is 45 ft tall casts a shadow that is 30 ft long. Nearby, Heather is walking her standard poodle, which casts a shadow that is 2.5 ft long. How tall is Heather's poodle?

a. 2.75 ft
b. 3.25 ft
c. 3.75 ft
d. 1.67 ft

394. A circular pool is filling with water. Assuming the water level will be 4 ft deep and the diameter is 20 ft, what is the volume of the water needed to fill the pool? ($\pi = 3.14$)

a. 251.2 ft^3
b. 1,256 ft^3
c. 5,024 ft^3
d. 3,140 ft^3

395. A cable is attached to a pole 36 ft above ground and fastened to a stake 15 ft from the base of the pole. In order to keep the pole perpendicular to the ground, how long is the cable?

a. 22 ft
b. 39 ft
c. 30 ft
d. 51 ft

396. Karen is buying a wallpaper border for her bedroom, which is 12 ft by 13 ft. If the border is sold in rolls of 5 yards each, how many rolls will she need to purchase?

a. 3
b. 4
c. 5
d. 6

397. The formula for the surface area of a sphere is $4\pi r^2$. What is the surface area of a ball with a diameter of 6 inches? Round to the nearest inch. ($\pi = 3.14$)

a. 452 in^2
b. 113 in^2
c. 38 in^2
d. 28 in^2

398. Brittney would like to carpet her bedroom. If her room is 11 ft by 14 ft, what is the area to be carpeted in square feet?

    a. 121 ft^2

    b. 25 ft^2

    c. 169 ft^2

    d. 154 ft^2

399. The scale on a map shows that 1 inch is equal to 14 miles. Shannon measured the distance on the map to be 17 inches. How far will she need to travel?

    a. 23.8 miles

    b. 238 miles

    c. 2,380 miles

    d. 23,800 miles

400. How far will a bowling ball roll in one rotation if the ball has a diameter of 10 inches? ($\pi = 3.14$)

    a. 31.4 in

    b. 78.5 in

    c. 15.7 in

    d. 62.8 in

401. A water sprinkler sprays in a circular pattern a distance of 10 ft. What is the circumference of the spray? ($\pi = 3.14$)

    a. 31.4 ft

    b. 314 ft

    c. 62.8 ft

    d. 628 ft

402. If a triangular sail has a vertical height of 83 ft and horizontal length of 40 ft, what is the area of the sail?

    a. 1,660 ft^2

    b. 1,155 ft^2

    c. 201 ft^2

    d. 3,320 ft^2

403. What is the volume of a ball whose radius is 4 inches? Round to the nearest inch. ($\pi = 3.14$)
  a. 201 in^3
  b. 268 in^3
  c. 804 in^3
  d. 33 in^3

404. If a tabletop has a diameter of 42 in, what is its surface area to the nearest inch? ($\pi = 3.14$)
  a. 1,384 in^2
  b. 1,319 in^2
  c. 1,385 in^2
  d. 5,539 in^2

405. An orange has a radius of 1.5 inches. Find the volume of one orange. ($\pi = 3.14$)
  a. 9.42 in^3
  b. 113.04 in^3
  c. 28.26 in^3
  d. 14.13 in^3

406. A fire and rescue squad places a 15 ft ladder against a burning building. If the ladder is 9 ft from the base of the building, how far up the building will the ladder reach?
  a. 8 ft
  b. 10 ft
  c. 12 ft
  d. 14 ft

407. Safe deposit boxes are rented at the bank. The dimensions of a box are 22 in by 5 in by 5 in. What is the volume of the box?
  a. 220 in^3
  b. 550 in^3
  c. 490 in^3
  d. 360 in^3

408. How many degrees does a minute hand move in 25 minutes?
  a. 25°
  b. 150°
  c. 60°
  d. 175°

409. Two planes leave the airport at the same time. Minutes later, plane A is 70 miles due north of the airport and plane B is 168 miles due east of the airport. How far apart are the two airplanes?

a. 182 miles

b. 119 miles

c. 163.8 miles

d. 238 miles

410. If the area of a small pizza is 78.5 in^2, what size pizza box would best fit the small pizza? (Note: Pizza boxes are measured according to the length of one side.)

a. 12 in

b. 11 in

c. 9 in

d. 10 in

411. Stuckeyburg is a small town in rural America. Use the map to approximate the area of the town.

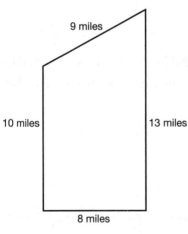

a. 40 miles^2

b. 104 miles^2

c. 93.5 miles^2

d. 92 miles^2

412. A rectangular field is to be fenced in completely. The width is 28 yd and the total area is 1,960 yd^2. What is the length of the field?

a. 1,932 yd

b. 70 yd

c. 31 yd

d. 473 yd

413. A circular print is being matted in a square frame. If the frame is 18 in by 18 in, and the radius of the print is 7 in, what is the area of the matting? (π = 3.14)

   a. 477.86 in²

   b. 170.14 in²

   c. 280.04 in²

   d. 288 in²

414. Ribbon is wrapped around a rectangular box that is 10 in by 8 in by 4 in. Using the illustration provided, determine how much ribbon is needed to wrap the box. Assume the amount of ribbon does not include a knot or bow.

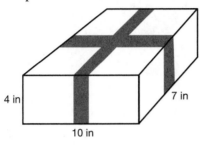

   a. 50 in

   b. 42 in

   c. 22 in

   d. 280 in

415. Pat is making a Christmas tree skirt. She needs to know how much fabric to buy. Using the illustration provided, determine the area of the skirt to the nearest foot.

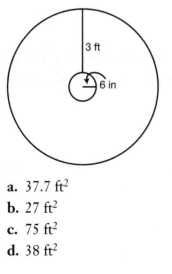

   a. 37.7 ft²

   b. 27 ft²

   c. 75 ft²

   d. 38 ft²

416. Mark intends to tile a kitchen floor, which is 9 ft by 11 ft. How many 6-inch tiles are needed to tile the floor?

a. 60

b. 99

c. 396

d. 449

417. A framed print measures 36 in by 22 in. If the print is enclosed by a 2-inch matting, what is the length of the diagonal of the print? Round to the nearest tenth. See illustration.

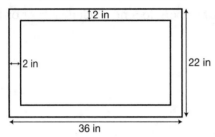

a. 36.7 in

b. 39.4 in

c. 26.5 in

d. 50 in

418. A 20-foot light post casts a shadow 25 feet long. At the same time, a building nearby casts a shadow 60 feet long. How tall is the building?

a. 48 ft

b. 75 ft

c. $8\frac{1}{3}$ ft

d. 95 ft

419. Barbara is wrapping a wedding gift that is contained within a rectangular box 20 in by 18 in by 4 in. How much wrapping paper will she need?

a. 512 in²

b. 1,440 in²

c. 1,024 in²

d. 92 in²

420. Mark is constructing a walkway around his inground pool. The pool is 20 ft by 40 ft and the walkway is intended to be 4 ft wide. What is the area of the walkway?

 a. 224 ft²

 b. 416 ft²

 c. 256 ft²

 d. 544 ft²

421. The picture frame shown below has outer dimensions of 8 in by 10 in and inner dimensions of 6 in by 8 in. Find the area of section A of the frame.

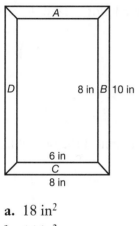

 a. 18 in²

 b. 14 in²

 c. 7 in²

 d. 9 in²

For questions 422 and 423, use the following illustration.

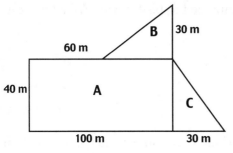

422. John is planning to purchase an irregularly shaped plot of land. Referring to the diagram, find the total area of the land.
  a. 6,400 m²
  b. 5,200 m²
  c. 4,500 m²
  d. 4,600 m²

423. Using the same illustration, determine the perimeter of the plot of land.
  a. 260 m
  b. 340 m
  c. 360 m
  d. 320 m

424. A weather vane is mounted on top of an 18 ft pole. If a 22 ft guy wire is staked to the ground to keep the pole perpendicular, how far is the stake from the base of the pole?
  a. 160 ft
  b. $\sqrt{808}$ ft
  c. 38 ft
  d. $\sqrt{160}$ or $4\sqrt{10}$ ft

425. A surveyor is hired to measure the width of a river. Using the illustration provided, determine the width of the river.

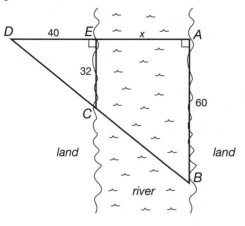

a. 48 ft
b. 8 ft
c. 35 ft
d. 75 ft

426. A publishing company is designing a book jacket for a newly published textbook. Find the area of the book jacket, given that the front cover is 8 in wide by 11 in high, the binding is 1.5 in by 11 in and the jacket will extend 2 inches inside the front and rear covers.

a. 236.5 in²
b. 192.5 in²
c. 188 in²
d. 232 in²

427. A Norman window is to be installed in a new home. Using the dimensions marked on the illustration, find the area of the window to the nearest tenth of an inch. ($\pi = 3.14$)

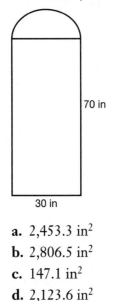

70 in

30 in

    a. 2,453.3 in²
    b. 2,806.5 in²
    c. 147.1 in²
    d. 2,123.6 in²

428. A surveyor is hired to measure the distance of the opening of a bay. Using the illustration and various measurements determined on land, find the distance of the opening of the bay.

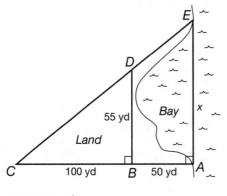

    a. 272.7 yds
    b. 82.5 yds
    c. 27.5 yds
    d. 205 yds

429. A car is initially 200 meters due west of a roundabout (traffic circle). If the car travels to the roundabout, continues halfway around the circle, exits due east, then travels an additional 160 meters, what is the total distance the car has traveled? Refer to diagram.

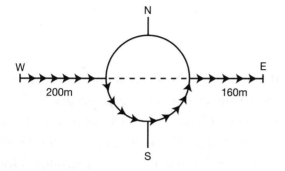

   a. 862.4 m

   b. 611.2 m

   c. 502.4 m

   d. 451.2 m

430. Steve Fossett is approaching the shores of Australia on the first successful solo hot air balloon ride around the world. His balloon, the Bud Light™ Spirit of Freedom, is being escorted by a boat (directly below him) that is 108 meters away. The boat is 144 meters from the shore. How far is Fossett's balloon from the shore?

   a. 252 m

   b. 95.2 m

   c. 126 m

   d. 180 m

431. Computer monitors are measured by their diagonals. If a monitor is advertised to be 17 in, what is the actual viewing area, assuming the screen is square? (Round to the nearest tenth.)

   a. 289 in^2

   b. 90.25 in^2

   c. 144.4 in^2

   d. 144.5 in^2

432. An elevated cylindrical shaped water tower is in need of paint. If the radius of the tower is 10 ft and the tower is 40 ft tall, what is the total area to be painted? ($\pi = 3.14$)

   a. 1,570 ft^2

   b. 2,826 ft^2

   c. 2,575 ft^2

   d. 3,140 ft^2

433. A sinking ship signals to the shore for help. Three individuals spot the signal from shore. The first individual is directly perpendicular to the sinking ship and 20 meters inland. The second individual is also 20 meters inland but 100 meters to the right of the first individual. The third is also 20 meters inland but 125 meters to the right of the first individual. How far off shore is the sinking ship? See illustration.

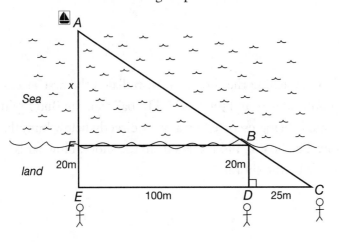

   a. 60 meters

   b. 136 meters

   c. 100 meters

   d. 80 meters

434. You are painting the surface of a silo that has a radius of 8 ft and height of 50 ft. What is the total surface area to be painted? Assume the top of the silo is $\frac{1}{2}$ a sphere and sets on the ground. Refer to the illustration.

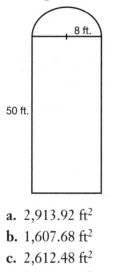

  a. 2,913.92 ft²
  b. 1,607.68 ft²
  c. 2,612.48 ft²
  d. 3,315.84 ft²

The Washington Monument is located in Washington D.C. Use the following illustration, which represents one of four identical sides, to answer questions 435 and 436.

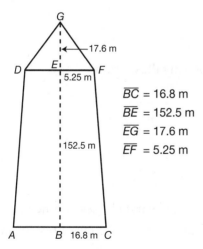

$\overline{BC}$ = 16.8 m
$\overline{BE}$ = 152.5 m
$\overline{EG}$ = 17.6 m
$\overline{EF}$ = 5.25 m

435. Find the height of the Washington Monument to the nearest tenth of a meter.
  a. 157.8 m
  b. 169.3 m
  c. 170.1 m
  d. 192.2 m

436. Find the surface area of the monument to the nearest meter.

    a. 13,820 m²

    b. 13,451 m²

    c. 3,455 m²

    d. 13,543 m²

437. An inground pool is filling with water. The shallow end is 3 ft deep and gradually slopes to the deepest end, which is 10 ft deep. The width of the pool is 15 ft and the length is 30 ft. What is the volume of the pool?

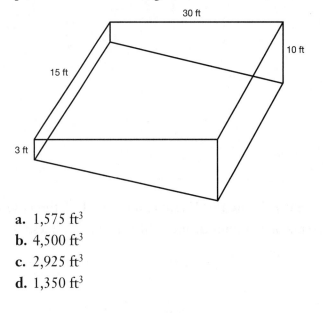

    a. 1,575 ft³

    b. 4,500 ft³

    c. 2,925 ft³

    d. 1,350 ft³

For questions 438 and 439, refer to the following illustration:

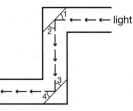

438. In a periscope, a pair of mirrors is mounted parallel to each other as shown. The path of light becomes a transversal. If ∠2 measures 50°, what is the measurement of ∠3?

    a. 50°

    b. 40°

    c. 130°

    d. 310°

439. Given that ∠2 measures 50°, what is the measurement of ∠4?

    a. 50°

    b. 40°

    c. 130°

    d. 85°

440. The angle measure of the base angles of an isosceles triangle are represented by x and the vertex angle is $3x + 20$. Find the measure of a base angle.

    a. 116°

    b. 40°

    c. 32°

    d. 14°

441. Using the information from question 440, find the measure of the vertex angle of the isosceles triangle.

    a. 32°

    b. 62°

    c. 58°

    d. 116°

442. In parallelogram $ABCD$, $\angle A = 5x + 2$ and $\angle C = 6x - 4$. Find the measure of $\angle A$.

    a. 32°

    b. 6°

    c. 84.7°

    d. 44°

443. The longer base of a trapezoid is three times the shorter base. The nonparallel sides are congruent. The nonparallel side is 5 cm more that the shorter base. The perimeter of the trapezoid is 40 cm. What is the length of the longer base?

    a. 15 cm

    b. 5 cm

    c. 10 cm

    d. 21 cm

444. The measure of the angles of a triangle are represented by $2x + 15$, $x + 20$, and $3x + 25$. Find the measure of the smallest angle within the triangle.

   a. 40°

   b. 85°

   c. 25°

   d. 55°

445. Suppose $ABCD$ is a rectangle. IF $\overline{AB}$ = 10 and $\overline{AD}$ = 6, find $\overline{BX}$ to the nearest tenth.

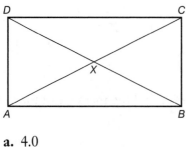

   a. 4.0

   b. 5.8

   c. 11.7

   d. 8.0

446. The perimeter of the parallelogram is 32 cm. What is the length of the longer side?

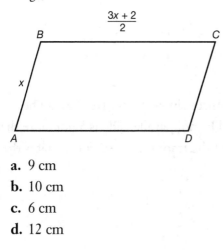

   a. 9 cm

   b. 10 cm

   c. 6 cm

   d. 12 cm

447. A door is 6 feet and 6 inches tall and 36 inches wide. What is the widest piece of sheetrock that can fit through the door? Round to the nearest inch.

   a. 114 in

   b. 86 in

   c. 85 in

   d. 69 in

448. The width of a rectangle is 20 cm. The diagonal is 8 cm more than the length. Find the length of the rectangle.

   a. 20

   b. 23

   c. 22

   d. 21

449. The measures of two complementary angles are in the ratio of 5:7. Find the measure of the smallest angle.

   a. 75°

   b. 37.5°

   c. 52.5°

   d. 105°

450. In parallelogram $ABCD$, $m\angle A = 3x + 10$ and $m\angle D = 2x + 30$, find the $m\angle A$.

   a. 70°

   b. 40°

   c. 86°

   d. 94°

451. Using the diagram below and the fact that $\angle A + \angle B + \angle C + \angle D = 320$, find $m\angle E$.

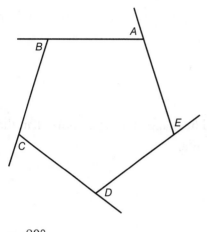

    a. 80°

    b. 40°

    c. 25°

    d. 75°

452. The base of a triangle is 4 times as long as its height. If together they measure 95 cm, what is the area of the triangle?

    a. 1,444 cm^2

    b. 100 cm^2

    c. 722 cm^2

    d. 95 cm^2

453. One method of finding the height of an object is to place a mirror on the ground and then position yourself so that the top of the object can be seen in the mirror. How high is a structure if a person who is 160 cm tall observes the top of a structure when the mirror is 100 m from the structure and the person is 8 m from the mirror?

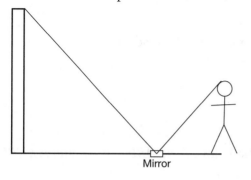

Mirror

   a. 50,000 cm
   b. 20,000 cm
   c. 2,000 cm
   d. 200 cm

454. Suppose *ABCD* is a parallelogram; ∠*B* = 120 and ∠2 = 40. Find *m*∠4.

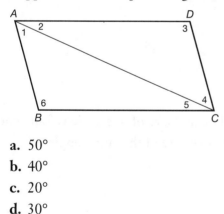

   a. 50°
   b. 40°
   c. 20°
   d. 30°

455. The length and width of a rectangle together measure 130 yards. Their difference is 8 yards. What is the area of the rectangle?

a. 4,209 yd²

b. 130 yd²

c. 3,233 yd²

d. 4,270 yd²

456. A sphere has a volume of 288π cm³. Find its radius.

a. 9.5 cm

b. 7 cm

c. 14 cm

d. 6 cm

457. Using the illustration provided below, if $m\angle ABE = 4x + 5$ and $m\angle CBD = 7x - 13$, find the measure of $\angle ABE$.

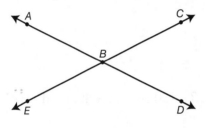

a. 151°

b. 73°

c. 107°

d. 29°

458. Two angles are complementary. The measure of one angle is four times the measure of the other. Find the measure of the larger angle.

a. 36°

b. 72°

c. 144°

d. 18°

459. If Gretta's bicycle has a 25-inch diameter wheel, how far will she travel in two turns of the wheel? (π = 3.14)

a. 491 in

b. 78.5 in

c. 100 in

d. 157 in

460. Two angles are supplementary. The measure of one is 30 more than twice the measure of the other. Find the measure of the larger angle.

a. 130°

b. 20°

c. 50°

d. 70°

461. Using the illustration provided, find the $m\angle AED$. Given $m\angle BEC = 5x - 36$ and $m\angle AED = 2x + 9$.

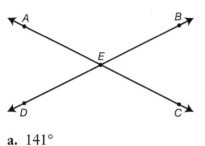

a. 141°

b. 69°

c. 111°

d. 39°

462. The measures of the angles of a triangle are in the ratio of 3:4:5. Find the measure of the largest angle.

a. 75°

b. 37.5°

c. 45°

d. 60°

463. A mailbox opening is 4.5 inches high and 5 inches wide. What is the widest piece of mail able to fit in the mailbox without bending? Round answer to the nearest tenth.

a. 9.5 inches

b. 2.2 inches

c. 6.7 inches

d. 8.9 inches

464. The figure below represents the cross section of a pipe $\frac{1}{2}$ inch thick that has an inside diameter of 3 inches. Find the area of the shaded region in terms of π.

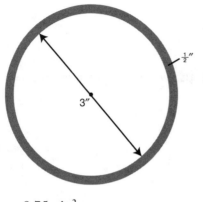

a. 8.75π in^2
b. 3.25π in^2
c. 7π in^2
d. 1.75π in^2

465. Using the same cross section of pipe from question 464, answer the following question. If the pipe is 18 inches long, what is the volume of the shaded region in terms of π?
a. 31.5π in^3
b. 126π in^3
c. 157.5 in^3
d. 58.5 in^3

466. A person travels 10 miles due north, 4 miles due west, 5 miles due north, and 12 miles due east. How far is that person from the starting point?
a. 23 miles northeast
b. 13 miles northeast
c. 17 miles northeast
d. 17 miles northwest

467. Using the illustration provided, find the area of the shaded region in terms of π.

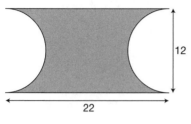

    a. $264 - 18\pi$
    b. $264 - 36\pi$
    c. $264 - 12\pi$
    d. $18\pi - 264$

468. Find how many square centimeters of paper are needed to create a label on a cylindrical can 45 cm tall with a circular base having diameter of 20 cm. Leave answer in terms of π.

    a. 450π cm^2
    b. $4,500\pi$ cm^2
    c. 900π cm^2
    d. $9,000\pi$ cm^2

469. Using the illustration provided below, if the measure $\angle AEB = 5x + 40$ and $\angle BEC = x + 20$, find $m\angle DEC$.

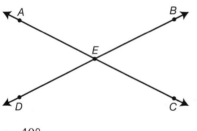

    a. $40°$
    b. $25°$
    c. $140°$
    d. $65°$

470. The structural support system for a bridge is shown in the illustration provided. $\overline{AD}$ is parallel to $\overline{BC}$, $\overline{BE}$ is parallel to $\overline{CD}$, and $\overline{AB}$ is parallel to $\overline{CF}$. Find $\angle CGE$.

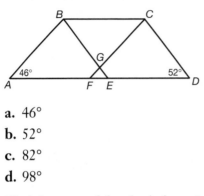

a. 46°

b. 52°

c. 82°

d. 98°

471. Find the area of the shaded portions, where $\overline{AB} = 6$ and $\overline{BC} = 10$. Leave answer in terms of π.

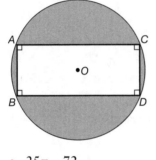

a. $25\pi - 72$

b. $25\pi - 48$

c. $25\pi - 8$

d. $100\pi - 48$

472. Find the area of the shaded region in terms of π.

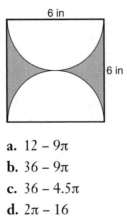

a. $12 - 9\pi$

b. $36 - 9\pi$

c. $36 - 4.5\pi$

d. $2\pi - 16$

473. On a piece of machinery, the centers of two pulleys are 3 feet apart, and the radius of each pulley is 6 inches. How long a belt (in feet) is needed to wrap around both pulleys?

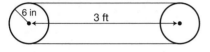

    a. $(6 + .5\pi)$ ft
    b. $(6 + .25\pi)$ ft
    c. $(6 + 12\pi)$ ft
    d. $(6 + \pi)$ ft

474. Find the measure of each angle of a regular 15-sided polygon to the nearest tenth.

    a. 24°
    b. 12°
    c. 128.6°
    d. 156°

475. A sand pile is shaped like a cone as illustrated below. How many cubic yards of sand are in the pile. Round to the nearest tenth. ($\pi = 3.14$)

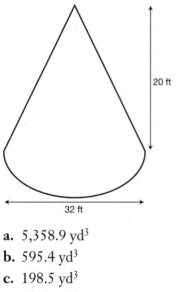

    a. 5,358.9 yd³
    b. 595.4 yd³
    c. 198.5 yd³
    d. 793.9 yd³

476. Find the area of the regular octagon with the following measurements.

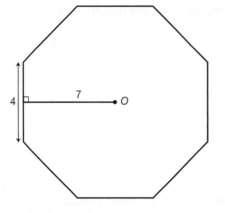

    a. 224 square units

    b. 112 square units

    c. 84 square units

    d. 169 square units

477. Two sides of a picture frame are glued together to form a corner. Each side is cut at a 45-degree angle. Using the illustration provided, find the measure of ∠A.

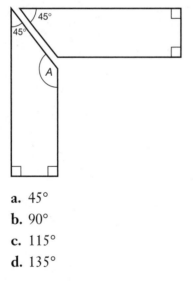

    a. 45°

    b. 90°

    c. 115°

    d. 135°

478. Find the total area of the shaded regions, if the radius of each circle is 5 cm. Leave answer in terms of π.

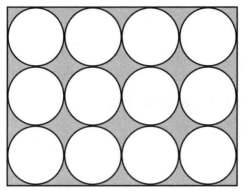

    a. 1,200 – 300π cm²

    b. 300 – 300π cm²

    c. 300π – 1,200 cm²

    d. 300π – 300 cm²

479. The road from town A to town B travels at a direction of N23°E. The road from town C to town D travels at a direction of S48°E. The roads intersect at location E. Find the measure of ∠BED, at the point of intersection.

    a. 71°

    b. 23°

    c. 109°

    d. 48°

480. The figure provided below represents a hexagonal-shaped nut. What is the measure of ∠*ABC*?

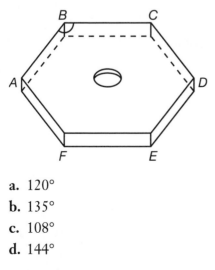

    a. 120°

    b. 135°

    c. 108°

    d. 144°

481. If the lengths of all sides of a box are doubled, how much is the volume increased?

   a. 2 times

   b. 4 times

   c. 6 times

   d. 8 times

482. If the radius of a circle is tripled, the circumference is

   a. multiplied by 3.

   b. multiplied by 6.

   c. multiplied by 9.

   d. multiplied by 12.

483. If the diameter of a sphere is doubled, the surface area is

   a. multiplied by 4.

   b. multiplied by 2.

   c. multiplied by 3.

   d. multiplied by 8.

484. If the diameter of a sphere is tripled, the volume is

   a. multiplied by 3.

   b. multiplied by 27.

   c. multiplied by 9.

   d. multiplied by 6.

485. If the radius of a cone is doubled, the volume is

   a. multiplied by 2.

   b. multiplied by 4.

   c. multiplied by 6.

   d. multiplied by 8.

486. If the radius of a cone is halved, the volume is

   a. multiplied by $\frac{1}{4}$.

   b. multiplied by $\frac{1}{2}$.

   c. multiplied by $\frac{1}{8}$.

   d. multiplied by $\frac{1}{16}$.

487. If the radius of a right cylinder is doubled and the height is halved, its volume

    a. remains the same.

    b. is multiplied by 2.

    c. is multiplied by 4.

    d. is multiplied by $\frac{1}{2}$.

488. If the radius of a right cylinder is doubled and the height is tripled, its volume is

    a. multiplied by 12.

    b. multiplied by 2.

    c. multiplied by 6

    d. multiplied by 3.

489. If each interior angle of a regular polygon has a measure of 150 degrees, how many sides does it have?

    a. 10

    b. 11

    c. 12

    d. 13

490. A box is 30 cm long, 8 cm wide and 12 cm high. How long is the diagonal $\overline{AB}$? Round to the nearest tenth.

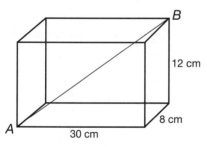

    a. 34.5 cm

    b. 32.1 cm

    c. 35.2 cm

    d. 33.3 cm

491. Find the area of the shaded region. Leave answer in terms of π.

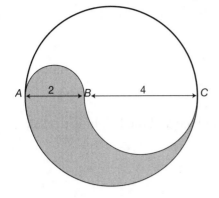

a. 16.5π

b. 30π

c. 3π

d. 7.5π

492. A round tower with a 40 meter circumference is surrounded by a security fence that is 8 meters from the tower. How long is the security fence in terms of π?

a. (40 + 16π) meters

b. (40 + 8π) meters

c. 48π meters

d. 56π meters

493. The figure below is two overlapping rectangles. Find the sum of ∠1 + ∠2 + ∠3 + ∠4.

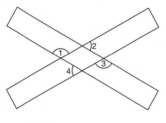

a. 360°

b. 90°

c. 180°

d. 540°

494. A solid is formed by cutting the top off of a cone with a slice parallel to the base, and then cutting a cylindrical hole into the resulting solid. Find the volume of the hollow solid in terms of π.

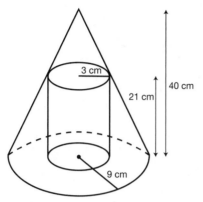

    a. 834π cm^3

    b. $2,880\pi$ cm^3

    c. 891π cm^3

    d. $1,326\pi$ cm^3

495. A rectangular container is 5 cm wide and 15 cm long, and contains water to a depth of 8 cm. An object is placed in the water and the water rises 2.3 cm. What is the volume of the object?

    a. 92 cm^3

    b. 276 cm^3

    c. 172.5 cm^3

    d. 312.5 cm^3

496. A concrete retaining wall is 120 feet long with ends shaped as shown. How many cubic yards of concrete are needed to construct the wall?

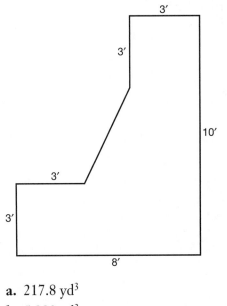

a. 217.8 yd³

b. 5,880 yd³

c. 653.3 yd³

d. 49 yd³

497. A spherical holding tank whose diameter to the outer surface is 20 feet is constructed of steel 1 inch thick. How many cubic feet of steel is needed to construct the holding tank? Round to the nearest integer value. (π = 3.14)

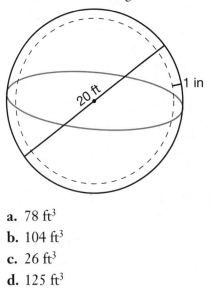

a. 78 ft³

b. 104 ft³

c. 26 ft³

d. 125 ft³

498. How many cubic inches of lead are there in the pencil? Round to the nearest thousandth. ($\pi = 3.14$)

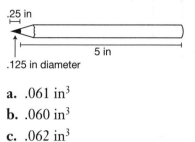

.25 in

5 in

.125 in diameter

a. .061 in^3
b. .060 in^3
c. .062 in^3
d. .063 in^3

499. A cylindrical hole with a diameter of 4 inches is cut through a cube. The edge of the cube is 5 inches. Find the volume of the hollowed solid in terms of π.

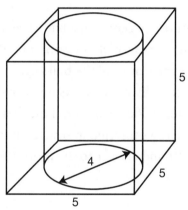

5

4

5

5

a. $125 - 80\pi$
b. $125 - 20\pi$
c. $80\pi - 125$
d. $20\pi - 125$

500. Find the area of the region.

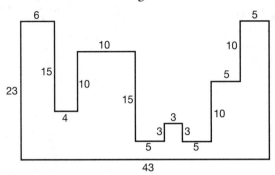

    a. 478 units^2

    b. 578 units^2

    c. 528 units^2

    d. 428 units^2

501. From a stationary point directly in front of the center of a bull's eye, Kim aims two arrows at the bull's eye. The first arrow nicks one point on the edge of the bull's eye; the second strikes the center of the bull's eye. Kim knows the second arrow traveled 20 meters since she knows how far she is from the target. If the bull's eye is 4 meters wide, how far did the first arrow travel? You may assume that the arrows traveled in straight-line paths and that the bull's eye is circular. Round answer to the nearest tenth.

    a. 19.9 meters

    b. 24 meters

    c. 22 meters

    d. 20.1 meters

Answer Explanations

The following explanations show one way in which each problem can be solved. You may have another method for solving these problems.

376. d. The area of a rectangle is *length × width.*

377. b. The volume of a sphere is $\frac{4}{3}$ times π times the radius cubed.

378. b. The area of a triangle is $\frac{1}{2}$ times the length of the base times the length of the height.

379. b. The surface area of a sphere is four times π times the radius squared.

380. d. The area of a circle is π times the radius squared.

381. a. The volume of a cylinder is π times the radius squared, times the height of the cylinder.

382. c. The perimeter of a square is four times the length of one side.

383. c. The area of the base is π times radius squared. The area of the curved region is two times π times radius times height. Notice there is only one circular region since the storage tank would be on the ground. This area would not be painted.

384. a. The area of a square is side squared or side times side.

385. c. The circumference or distance around a circle is π times the diameter.

386. a. The perimeter of a rectangle is two times the length plus two times the width.

387. d. The volume of a cube is the length of the side cubed or the length of the side times the length of the side times the length of the side.

388. a. The perimeter of a triangle is length of side *a* plus length of side *b* plus length of side *c*.

389. c. The area of a rectangle is length times width. Therefore, the area of the racquetball court is equal to 40 ft times 20 ft or 800 ft². If you chose answer **d**, you found the perimeter or distance around the court.

390. a. The width of the field, 180 ft, must be divided by the width of the mower, 2 ft. The result is that he must mow across the lawn 90 times. If you chose **b**, you calculated as if he were mowing the length of the field. If you chose **c**, you combined length and width, which would result in mowing the field twice.

391. a. The area of the room is the sum of the area of four rectangular walls. Each wall has an area of length times width, or (9)(5.5), which equals 49.5 ft². Multiply this by 4 which equals 198 ft². If you chose **b**, you added 9 ft and 5.5 ft instead of multiplying.

392. c. The ceiling fan follows a circular pattern, therefore area = πr^2. Area = $(3.14)(25)^2$ = 1,962.5 in². If you chose **a**, the incorrect formula you used was $\pi^2 r$. If you chose **d**, the incorrect formula you used was πd.

393. c. To find the height of Heather's poodle, set up the following proportion: height of the building/shadow of the building = height of the poodle/shadow of the poodle or $\frac{45}{30} = \frac{x}{2.5}$. Cross-multiply, 112.5 = 30x. Solve for x; 3.75 = x. If you chose **d**, the proportion was set up incorrectly as $\frac{45}{30} = \frac{2.5}{x}$.

394. b. The volume of a cylinder is $\pi r^2 h$. Using a height of 4 ft and radius of 10 ft, the volume of the pool is $(3.14)(10)^2(4)$ or 1,256 ft³. If you chose **a**, you used πdh instead of $\pi r^2 h$. If you chose **c**, you used the diameter squared instead of the radius squared.

395. b. The connection of the pole with the ground forms the right angle of a triangle. The length of the pole is a leg within the right triangle. The distance between the stake and the pole is also a leg within the right triangle. The question is to find the length of the cable, which is the hypotenuse. Using the Pythagorean theorem: $(36)^2 + (15)^2 = c^2$; 1,296 + 225 = c^2; 1,521 = c^2; 39 = c.

396. b. The distance around the room is 2(12) + 2(13) or 50 ft. Each roll of border is 5(3) or 15 ft. By dividing the total distance, 50 ft, by the length of each roll, 15 ft, we find we need 3.33 rolls. Since a roll cannot be subdivided, 4 rolls will be needed.

397. b. If the diameter of a sphere is 6 inches, the radius is 3 inches. The radius of a circle is half the diameter. Using the radius of 3 inches, surface area equals $(4)(3.14)(3)^2$ or 113.04 in². Rounding this to the nearest inch is 113 in². If you chose **a**, you used the diameter rather than the radius. If you chose **c**, you did not square the radius. If you chose **d**, you omitted the value 4 from the formula for the surface area of a sphere.

398. d. The area of a rectangle is length times width. Using the dimensions described, area = (11)(14) or 154 ft².

399. b. To find how far Shannon will travel, set up the following proportion: $\frac{1 \text{ inch}}{14 \text{ miles}} = \frac{17 \text{ inches}}{x \text{ miles}}$. Cross multiply, $x = 238$ miles.

400. a. The circumference of a circle is πd. Using the diameter of 10 inches, the circumference is equal to $(3.14)(10)$ or 31.4 inches. If you chose **b**, you found the area of a circle. If you chose **c**, you mistakenly used πr for circumference rather than $2\pi r$. If you chose **d**, you used the diameter rather than the radius.

401. c. The circumference of a circle is πd. Since 10 ft represents the radius, the diameter is 20 feet. The diameter of a circle is twice the radius. Therefore, the circumference is $(3.14)(20)$ or 62.8 ft. If you chose **a**, you used πr rather than $2\pi r$. If you chose **b**, you found the area rather than circumference.

402. a. The area of a triangle is $\frac{1}{2}$(base)(height). Using the dimensions given, area $= \frac{1}{2}(40)(83)$ or 1,660 ft^2. If you chose **d**, you omitted $\frac{1}{2}$ from the formula.

403. b. The volume of a sphere is $\frac{4}{3}\pi r^3$. Using the dimensions given, volume $= \frac{4}{3}(3.14)(4)^3$ or 267.9. Rounding this answer to the nearest inch is 268 in^3. If you chose **a**, you found the surface area rather than volume. If you chose **c**, you miscalculated surface area by using the diameter.

404. c. The area of a circle is πr^2. The diameter = 42 in, radius = 42 ÷ 2 = 21 in, so $(3.14)(21)^2 = 1,384.74$ in^2. Rounding to the nearest inch, the answer is 1,385 in^2. If you chose **a**, you rounded the final answer incorrectly. If you chose **d**, you used the diameter rather than the radius.

405. d. To find the volume of a sphere, use the formula Volume $= \frac{4}{3}\pi r^3$. Volume $= \frac{4}{3}(3.14)(1.5)^3 = 14.13$ in^3. If you chose **a**, you squared the radius instead of cubing the radius. If you chose **b**, you cubed the diameter instead of the radius. If you chose **c**, you found the surface area of the sphere, not the volume.

406. c. The ladder forms a right triangle with the building. The length of the ladder is the hypotenuse and the distance from the base of the building is a leg. The question asks you to solve for the remaining leg of the triangle, or how far up the building the ladder will reach. Using the Pythagorean theorem: $9^2 + b^2 = 15^2$; $81 + b^2 = 225$; $81 + b^2 - 81 = 225 - 81$; $b^2 = 144$; $b = 12$.

407. b. The volume of a rectangular solid is length times width times depth. Using the dimensions in the question, volume = (22)(5)(5) or 550 in³. If you chose **c**, you found the surface area of the box.

408. b. A minute hand moves 180 degrees in 30 minutes. Using the following proportion: $\frac{30 \text{ minutes}}{180 \text{ degrees}} = \frac{25 \text{ minutes}}{x \text{ degrees}}$. Cross-multiply, $30x = 4,500$. Solve for x; $x = 150$ degrees.

409. a. The planes are traveling perpendicular to each other. The course they are traveling forms the legs of a right triangle. The question requires us to find the distance between the planes or the length of the hypotenuse. Using the Pythagorean theorem $70^2 + 168^2 = c^2$; $4,900 + 28,224 = c^2$; $33,124 = c^2$; $c = 182$ miles. If you chose **c**, you assigned the hypotenuse the value of 168 miles and solved for a leg rather than the hypotenuse. If you chose **d**, you added the legs together rather than using the Pythagorean theorem.

410. d. The area of a small pizza is 78.5 in². The question requires us to find the diameter of the pizza in order to select the most appropriate box. Area is equal to πr^2. Therefore, $78.5 = \pi r^2$; divide by π (3.14); $78.5 \div 3.14 = \pi r^2 \div 3.14$; $25 = r^2$; $5 = r$. The diameter is twice the radius or 10 inches. Therefore, the box is also 10 inches.

411. d. The area of Stuckeyburg can be found by dividing the region into a rectangle and a triangle. Find the area of the rectangle ($A = lw$) and add the area of the triangle ($\frac{1}{2}bh$) for the total area of the region. Referring to the diagram, the area of the rectangle is $(10)(8) = 80$ miles². The area of the triangle is $\frac{1}{2}(8)(3) = 12$ miles². The sum of the two regions is 80 miles² + 12 miles² = 92 miles². If you chose **a**, you found the perimeter. If you chose **b**, you found the area of the rectangular region but did not include the triangular region.

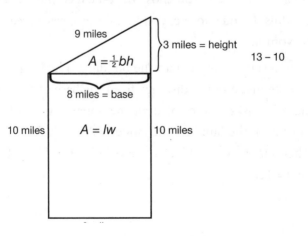

412. b. The area of a rectangle is length times width. Using the formula 1,960 yd^2 = (l)(28), solve for l by dividing both sides by 28; l = 70 yards.

413. b. To find the area of the matting, subtract the area of the print from the area of the frame. The area of the print is found using πr^2 or (3.14)(7)^2 which equals 153.86 in^2. The area of the frame is length of side times length of side or (18)(18), which equals 324 in^2. The difference, 324 in^2 – 153.86 in^2 or 170.14 in^2, is the area of the matting. If you chose **c**, you mistakenly used the formula for the circumference of a circle, $2\pi r$, instead of the area of a circle, πr^2.

414. a. The ribbon will travel the length (10 in) twice, the width (7 in) twice and the height (4 in) four times. Adding up these distances will determine the total amount of ribbon needed. 10 in + 10 in + 7 in + 7 in + 4 in + 4 in + 4 in + 4 in = 50 inches of ribbon. If you chose **b**, you missed two sides of 4 inches. If you chose **d**, you calculated the volume of the box.

415. d. To find the area of the skirt, find the area of the outer circle minus the area of the inner circle. The area of the outer circle is π (3.5)^2 or 38.465 in^2. The area of the inner circle is π (.5)^2 or .785 in^2. The difference is 38.465 – .785 or 37.68 ft^2. The answer, rounded to the nearest foot, is 38 ft^2. If you chose **a**, you rounded to the nearest tenth of a foot. If you chose **b**, you miscalculated the radius of the outer circle as being 3 feet instead of 3.5 feet.

416. c. Since the tiles are measured in inches, convert the area of the floor to inches as well. The length of the floor is 9 ft × 12 in per foot = 108 in. The width of the floor is 11 ft × 12 in per foot = 132 in. The formula for area of a rectangle is length × width. Therefore, the area of the kitchen floor is 108 in × 132 in or 14,256 in^2. The area of one tile is 6 in × 6 in or 36 in^2. Finally, divide the total number of square inches by 36 in^2 or 14,256 in^2 divided by 36 in^2 = 396 tiles.

417. a. If a framed print is enclosed by a 2-inch matting, the print is 4 in less in length and height. Therefore, the picture is 32 in by 18 in. These measurements along with the diagonal form a right triangle. Using the Pythagorean theorem, solve for the diagonal. 32^2 + 18^2 = c^2; 1,024 + 324 = c^2; 1,348 = c^2; 36.7 = c. If you chose **b**, you reduced the print 2 inches less than the frame in length and height rather than 4 inches.

418. a. To find the height of the building set up the following proportion: $\frac{\text{height of the light post}}{\text{shadow of light post}} = \frac{\text{height of the building}}{\text{shadow of building}}$ or $\frac{20}{25} = \frac{x}{60}$. Cross-multiply: $1{,}200 = 25x$. Solve for x by dividing both sides by 25; $x = 48$. If you chose **b**, you set up the proportion incorrectly as $\frac{20}{25} = \frac{60}{x}$. If you chose **c**, you set up the proportion incorrectly as $\frac{60}{25} = \frac{20}{x}$.

419. c. The surface area of the box is the sum of the areas of all six sides. Two sides are 20 in by 18 in or $(20)(18) = 360$ in^2. Two sides are 18 in by 4 in or $(18)(4) = 72$ in^2. The last two sides are 20 in by 4 in or $(20)(4) = 80$ in^2. Adding up all six sides: 360 in^2 + 360 in^2 + 77 in^2 + 77 in^2 + 80 in^2 + 80 in^2 = 1,024 in^2, is the total area. If you chose **a**, you did not double all sides. If you chose **b**, you calculated the volume of the box.

420. d. The area of the walkway is equal to the entire area (area of the walkway and pool) minus the area of the pool. The area of the entire region is length times width. Since the pool is 20 feet wide and the walkway adds 4 feet onto each side, the width of the rectangle formed by the walkway and pool is 20 + 4 + 4 = 28 feet. Since the pool is 40 feet long and the walkway adds 4 feet onto each side, the length of the rectangle formed by the walkway and pool is 40 + 4 + 4 = 48 feet. Therefore, the area of the walkway and pool is $(28)(48) = 1{,}344$ ft^2. The area of the pool is $(20)(40) = 800$ ft^2. 1,344 ft^2 – 800 ft^2 = 544 ft^2. If you chose **c**, you extended the entire area's length and width by 4 feet instead of 8 feet.

421. c. The area described as section A is a trapezoid. The formula for the area of a trapezoid is $\frac{1}{2}h(b_1 + b_2)$. The height of the trapezoid is 1 inch, b_1 is 6 inches, and b_2 is 8 inches. Using these dimensions, area = $\frac{1}{2}(1)(6 + 8)$ or 7 in^2. If you chose **b**, you used a height of 2 inches rather than 1 inch. If you chose **d**, you found the area of section B or D.

422. b. To find the total area, add the area of region A plus the area of region B plus the area of region C. The area of region A is length times width or $(100)(40) = 4{,}000$ m^2. Area of region B is $\frac{1}{2}bh$ or $\frac{1}{2}(40)(30) = 600$ m^2. The area of region C is $\frac{1}{2}bh$ or $\frac{1}{2}(30)(40) = 600$ m^2. The combined area is the sum of the previous areas or 4,000 + 600 + 600 = 5,200 m^2. If you chose **a**, you miscalculated the area of a triangle as bh instead of $\frac{1}{2}bh$. If you chose **c**, you found only the area of the rectangle. If you chose **d**, you found the area of the rectangle and only one of the triangles.

423. **c.** To find the perimeter, we must know the length of all sides. According to the diagram, we must find the length of the hypotenuse for the triangular regions B and C. Using the Pythagorean theorem for triangular region B, $30^2 + 40^2 = c^2$; $900 + 1,600 = c^2$; $2,500 = c^2$; 50 m $= c$. The hypotenuse for triangular region C is also 50 m since the legs are 30 m and 40 m as well. Now adding the length of all sides, 40 m + 100 m + 30 m + 50 m + 30 m + 50 m + 60 m = 360 m, the perimeter of the plot of land. If you chose **a**, you did not calculate in the hypotenuse on either triangle. If you chose **b**, you miscalculated the hypotenuse as having a length of 40 m. If you chose **d**, you miscalculated the hypotenuse as having a length of 30 m.

424. **d.** The 18 ft pole is perpendicular to the ground forming the right angle of a triangle. The 22 ft guy wire represents the hypotenuse. The task is to find the length of the remaining leg in the triangle. Using the Pythagorean theorem: $18^2 + b^2 = 22^2$; $324 + b^2 = 484$; $b^2 = 160$; $b = \sqrt{160}$ or $4\sqrt{10}$. If you chose **a**, you did not take the square root.

425. **c.** $\triangle ABD$ is similar to $\triangle ECD$. Using this fact, the following proportion is true: $\frac{DE}{EC} = \frac{DA}{AB}$ or $\frac{40}{32} = \frac{(40 + x)}{60}$. Cross-multiply, $2,400 = 32(40 + x)$; $2,400 = 1,280 + 32x$. Subtract 1,280; $1,120 = 32x$; divide by 32; $x = 35$ feet.

426. **a.** The area of the front cover is length times width or $(8)(11) = 88$ in^2. The rear cover is the same as the front, 88 in^2. The area of the binding is length times width or $(1.5)(11) = 16.5$ in^2. The extension inside the front cover is length times width or $(2)(11) = 22$ in^2. The extension inside the rear cover is also 22 in^2. The total area is the sum of all previous areas or 88 in^2 + 88 in^2 + 16.5 in^2 + 22 in^2 + 22 in^2 or 236.5 in^2. If you chose **b**, you did not calculate the extensions inside the front and rear covers. If you chose **c**, you miscalculated the area of the binding as $(1.5)(8)$ and omitted the extensions inside the front and rear covers. If you chose **d**, you miscalculated the area of the binding as $(1.5)(8)$ only.

427. **a.** To find the area of the rectangular region, multiply length times width or $(30)(70)$, which equals 2,100 in^2. To find the area of the semi-circle, multiply $\frac{1}{2}$ times πr^2 or $\frac{1}{2}\pi(15)^2$ which equals 353.25 in^2. Add the two areas together, 2,100 plus 353.25 or 2,453.3, rounded to the nearest tenth, for the area of the entire window. If you chose **b**, you included the area of a circle, not a semi-circle.

428. b. $\triangle ACE$ and $\triangle BCD$ are similar triangles. Using this fact, the following proportion is true: $\frac{\overline{CB}}{\overline{BD}} = \frac{\overline{CA}}{\overline{AE}}$ or $\frac{100}{55} = \frac{150}{x}$. Cross-multiply, $100x = 8{,}250$. Divide by 100 to solve for x; $x = 82.5$ yards. If you chose **a** or **c**, you set up the proportion incorrectly.

429. b. The question requires us to find the distance around the semi-circle. This distance will then be added to the distance traveled before entering the roundabout, 200 m, and the distance traveled after exiting the roundabout, 160 m. According to the diagram, the diameter of the roundabout is 160 m. The distance or circumference of half a circle is $\frac{1}{2}\pi d$, $\frac{1}{2}(3.14)(160)$ or 251.2 m. The total distance or sum is 200 m + 160 m + 251.2 m = 611.2 m. If you chose **a**, you included the distance around the entire circle. If you chose **c**, you found the distance around the circle. If you chose **d**, you did not include the distance after exiting the circle, 160 m.

430. d. The boat is the triangle's right angle. The distance between the balloon and the boat is 108 meters, one leg. The distance between the boat and the land, 144 meters, is the second leg. The distance between the balloon and the land, which is what we are finding, is the hypotenuse. Using the Pythagorean theorem: $108^2 + 144^2 = c^2$; $11{,}664 + 20{,}736 = c^2$; $32{,}400 = c^2$; $c = 180$ m.

431. d. Since the monitor is square, the diagonal and length of the sides of the monitor form an isosceles right triangle. The question requires one to find the length of one leg to find the area. Using the Pythagorean theorem: $s^2 + s^2 = 17^2$; $2s^2 = 289$. Divide by 2; $s^2 = 144.5$. The area of the square monitor is s^2; thus the area is 144.5 in². If you chose **a**, you simply squared the diagonal or which does not give you the area of the monitor.

432. b. To find the surface area of a cylinder, use the following formula: surface area = $2\pi r^2 + \pi dh$. Therefore, the surface area = $2(3.14)(10)^2 + (3.14)(20)(40)$ or 3,140 ft². If you chose **b**, you found the surface area of the circular top and forgot about the bottom of the water tower. However, the bottom of the tower would need painting since the tank is elevated.

433. d. Using the concept of similar triangles, $\triangle CDB$ is similar to $\triangle CEA$, so set up the following proportion: $\frac{25}{20} = \frac{125}{(x+20)}$. Cross-multiply, $25x + 500 = 2{,}500$. Subtract 500; $25x = 2{,}000$; Divide by 25; $x = 80$. If you chose **b**, the proportion was set up incorrectly as $\frac{25}{20} = \frac{(x+20)}{125}$.

434. a. To find the total surface area of the silo, add the surface area of the cylinder to the surface area of $\frac{1}{2}$ of the sphere. To find the area of the cylinder, use the formula πhd or $(3.14)(50)(16)$ which equals $2{,}512$ ft². The area of $\frac{1}{2}$ a sphere is $(\frac{1}{2})(4)\pi r^2$. Using a radius of 8 ft, the area is $(\frac{1}{2})(4)\pi(8)^2 = 401.92$ ft². Adding the area of the cylinder plus $\frac{1}{2}$ the sphere is $2{,}512 + 401.92 = 2{,}913.92$ ft². If you chose **b**, your miscalculation was in finding the area of $\frac{1}{2}$ the sphere. You used the diameter rather than the radius. If you chose **d**, you found the surface area of the entire sphere, not just half.

435. c. The height of the monument is the sum of $\overline{BE}$ plus $\overline{EG}$. Therefore, the height is $152.5 + 17.6$ or 170.1 meters. If you chose **a**, you added $\overline{BE} + \overline{EF}$. If you chose **b**, you added $\overline{BE} + \overline{BC}$.

436. a. The surface area of the monument is the sum of 4 sides of a trapezoidal shape plus 4 sides of a triangular shape. The trapezoid $DFCA$ has a height of 152.5m $(\overline{BE})$, $b_1 = 33.6$ $(\overline{AC})$, and $b_2 = 10.5$ $(\overline{DF})$. The area is $\frac{1}{2}h(b_1 + b_2)$ or $\frac{1}{2}(152.5)(33.6 + 10.5)$ which equals $3{,}362.625$ m². The triangle DGF has $b = 10.5$ and $h = 17.6$. The area is $\frac{1}{2}bh$ or $\frac{1}{2}(10.5)(17.6)$ which equals 92.4 m². The sum of 4 trapezoidal regions, $(4)(3{,}362.625) = 13{,}450.5$ m², plus 4 triangular regions, $4(92.4) = 369.6$ m², is $13{,}820.1$ m². Rounding this answer to the nearest meter is $13{,}820$ m². If you chose **b**, you found the area of the trapezoidal regions only. If you chose **c**, you found the area of one trapezoidal region and one triangular region. If you chose **d**, you found the area of 4 trapezoidal regions and one triangular region.

437. **c.** The volume of a rectangular solid is length times width times height. First, calculate what the volume would be if the entire pool had a depth of 10 ft. The volume would be (10)(30)(15) or 4,500 ft³. Now subtract the area under the sloped plane, a triangular solid. The volume of the region is $\frac{1}{2}$(*base*)(*height*)(*depth*) or $\frac{1}{2}$(7)(30)(15) or 1,575 ft³. Subtract: 4,500 ft³ minus 1,575 ft³ results in 2,925 ft³ as the volume of the pool. If you chose **a**, this is the volume of the triangular solid under the sloped plane in the pool. If you chose **b**, you did not calculate the slope of the pool, but rather a pool that is consistently 10 feet deep.

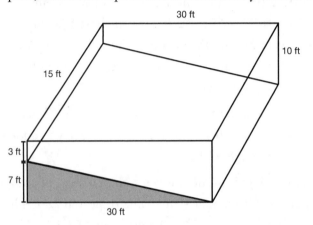

438. **a.** Two parallel lines cut by a transversal form alternate interior angles that are congruent. The two parallel lines are formed by the mirrors, and the path of light is the transversal. Therefore, ∠2 and ∠3 are alternate interior angles that are congruent. If ∠2 measures 50°, ∠3 is also 50°. If you chose **b**, your mistake was assuming ∠2 and ∠3 are complementary angles. If you chose **c**, your mistake was assuming ∠2 and ∠3 are supplementary angles.

439. **b.** Knowing that ∠4 + ∠3 + the right angle placed between ∠4 and ∠3, equals 180 and the fact that ∠3 = 50, we simply subtract 180 – 90 – 50, which equals 40. If you chose **a**, you assumed that ∠3 and ∠4 are vertical angles. If you chose **c**, you assumed that ∠3 and ∠4 are supplementary.

440. **c.** The sum of the measures of the angles of a triangle is 180. The question is asking us to solve for x. The equation is $x + x + 3x + 20 = 180$. Simplifying the equation, $5x + 20 = 180$. Subtract 20 from each side; $5x = 160$. Divide each side by 5; $x = 32$. If you chose **a**, you solved for the vertex angle. If you chose **b**, you wrote the original equation incorrectly as $x + 3x + 20 = 180$. If you chose **d**, you wrote the original equation incorrectly as $x + x + 3x + 10 = 90$.

441. **d.** Since we solved for x in the previous question, simply substitute $x = 32$ into the equation for the vertex angle, $3x + 20$. The result is $116°$. If you chose **a**, you solved for the base angle. If you chose **b**, the original equation was written incorrectly as $x + x + 3x + 20 = 90$.

442. **a.** Opposite angles of a parallelogram are equal in measure. Using this fact, $\angle A = \angle C$ or $5x + 2 = 6x - 4$. Subtract $5x$ from both sides; $2 = x - 4$. Add 4 to both sides; $6 = x$. Now substitute $x = 6$ into the expression for $\angle A$; $6(6) - 4 = 36 - 4$ or 32. If you chose **b**, you solved for x, not the angle. If you chose **c**, you assumed the angles were supplementary. If you chose **d**, you assumed the angles were complementary.

443. **a.** The two bases of the trapezoid are represented by x and $3x$. The nonparallel sides are each $x + 5$. Setting up the equation for the perimeter will allow us to solve for x; $x + 3x + x + 5 + x + 5 = 40$. Simplify to $6x + 10 = 40$. Subtract 10 from both sides; $6x = 30$. Divide both sides by 6; $x = 5$. The longer base is represented by $3x$. Using substitution, $3x$ or $(3)(5)$ equals 15, the longer base. If you chose **b**, you solved for the shorter base. If you chose **c**, you solved for the nonparallel side. If you chose **d**, the original equation was incorrect, $x + x + 5 + 3x = 40$.

444. **a.** The sum of the measures of the angles of a triangle is 180. Using this information, we can write the equation $2x + 15 + x + 20 + 3x + 25 = 180$. Simplify the equation; $6x + 60 = 180$. Subtract 60 from both sides; $6x = 120$. Divide both sides by 6; $x = 20$. Now substitute 20 for x in each expression to find the smallest angle. The smallest angle is found using the expression $x + 20$; $20 + 20 = 40$. If you chose **b**, this was the largest angle within the triangle. If you chose **c**, the original equation was incorrectly written as $2x + 15 + x + 20 + 3x + 25 = 90$. If you chose **d**, this was the angle that lies numerically between the smallest and largest angle measurements.

445. b. $\overline{AB}$ and $\overline{AD}$ are the legs of a right triangle. $\overline{DB}$ is the hypotenuse and $\overline{BX}$ is equal to $\frac{1}{2}$ of $\overline{DB}$. Solving for the hypotenuse, we use the Pythagorean theorem, $a^2 + b^2 = c^2$; $10^2 + 6^2 = \overline{DB}^2$; $100 + 36 = \overline{DB}^2$; $136 = \overline{DB}^2$. $\overline{DB} = 11.66$; $\frac{1}{2}$ of $\overline{DB} = 5.8$. If you chose **a**, you assigned 10 as the length of the hypotenuse. If you chose **d**, the initial error was the same as choice **a**. In addition, you solved for $\overline{DB}$ and not $\frac{1}{2}\overline{DB}$.

446. b. The perimeter of a parallelogram is the sum of the lengths of all four sides. Using this information and the fact that opposite sides of a parallelogram are equal, we can write the following equation: $x + x + \frac{3x + 2}{2} + \frac{(3x + 2)}{2} = 32$. Simplify to $2x + 3x + 2 = 32$. Simplify again; $5x + 2 = 32$. Subtract 2 from both sides; $5x = 30$. Divide both sides by 5; $x = 6$. The longer base is represented by $\frac{(3x + 2)}{2}$. Using substitution, $\frac{(3(6) + 2)}{2}$ equals 10. If you chose **c**, you solved for the shorter side.

447. b. To find the width of the piece of sheetrock that can fit through the door, we recognize it to be equal to the length of the diagonal of the door frame. If the height of the door is 6 ft 6 in, this is equivalent to 78 inches. Using the Pythagorean theorem, $a = 78$ and $b = 36$, we will solve for c. $(78)^2 + (36)^2 = c^2$. Simplify: $6{,}084 + 1{,}296 = c^2$; $7{,}380 = c^2$. Take the square root of both sides, $c = 86$. If you chose **a**, you added $78 + 36$. If you chose **c**, you rounded incorrectly. If you chose **d**, you assigned 78 inches as the hypotenuse, c.

448. d. To find the length of the rectangle, we will use the Pythagorean theorem. The width, a, is 20. The diagonal, c, is $x + 8$. The length, b, is x; $a^2 + b^2 = c^2$; $20^2 + x^2 = (x + 8)^2$. After multiplying the two binomials (using FOIL), $400 + x^2 = x^2 + 16x + 64$. Subtract x^2 from both sides; $400 = 16x + 64$. Subtract 64 from both sides; $336 = 16x$. Divide both sides by 16; $21 = x$. If you chose **a**, you incorrectly determined the diagonal to be 28.

449. b. Two angles are complementary if their sum is $90°$. Using this fact, we can establish the following equation: $5x + 7x = 90$. Simplify; $12x = 90$. Divide both sides of the equation by 12; $x = 7.5$. The smallest angle is represented by $5x$. Therefore $5x = 5(7.5)$ or 37.5, the smallest angle measurement. If you chose **a**, the original equation was set equal to 180 rather than 90. If you chose **c**, you solved for the largest angle. If you chose **d**, the original equation was set equal to 180 and you solved for the largest angle as well.

450. d. Adjacent angles in a parallelogram are supplementary. $\angle A$ and $\angle D$ are adjacent angles. Therefore, $\angle A + \angle D = 180$; $3x + 10 + 2x + 30 = 180$. Simplifying, $5x + 40 = 180$. Subtract 40 from both sides, $5x = 140$. Divide both sides by 5; $x = 28$. $\angle A = 3x + 10$ or $3(28) + 10$ which equals 94. If you chose **a**, you assumed $\angle A = \angle D$. If you chose **b**, you assumed $\angle A + \angle D = 90$. If you chose **c**, you solved for $\angle D$ instead of $\angle A$.

451. b. The sum of the measures of the exterior angles of any polygon is 360°. Therefore, if the sum of four of the five angles equals 320, to find the fifth simply subtract 320 from 360, which equals 40. If you chose **a**, you divided 325 by 4, assuming all four angles are equal in measure and assigned this value to the fifth angle, $\angle E$.

452. c. This problem requires two steps. First, determine the base and height of the triangle. Second, determine the area of the triangle. To determine the base and height we will use the equation $x + 4x = 95$. Simplifying, $5x = 95$. Divide both sides by 5, $x = 19$. By substitution, the height is 19 and the base is $4(19)$ or 76. The area of the triangle is found by using the formula $area = \frac{1}{2} base \times height$. Therefore, the area $= \frac{1}{2}(76)(19)$ or 722 cm^2. If you chose **a**, the area formula was incorrect. Area $= \frac{1}{2} base \times height$, not $base \times height$. If you chose **b**, the original equation $x + 4x = 95$ was simplified incorrectly as $4x^2 = 95$.

453. c. To solve for the height of the structure, solve the following proportion: $\frac{x}{100} = \frac{160}{8}$. Cross-multiply, $8x = 16,000$. Divide both sides by 8; $x = 2,000$. If you chose **b** or **d**, you made a decimal error.

454. c. In parallelogram $ABCD$, $\angle 2$ is equal in measurement to $\angle 5$. $\angle 2$ and $\angle 5$ are alternate interior angles, which are congruent. If $\angle B$ is 120, then $\angle B + \angle 5 + \angle 4 = 180$. Adjacent angles in a parallelogram are supplementary. Therefore, $40 + 120 + x = 180$. Simplifying, $160 + x = 180$. Subtract 160 from both sides; $x = 20$. If you chose **a**, you assumed $\angle 4 + \angle 5 = 90$. If you chose **d**, you assumed $\angle 4$ is $\frac{1}{2}(\angle 4 + \angle 5)$.

455. a. There are two ways of solving this problem. The first method requires a linear equation with one variable. The second method requires a system of equations with two variables. Let the length of the rectangle equal x. Let the width of the rectangle equal $x + 8$. Together they measure 130 yards. Therefore, $x + x + 8 = 130$. Simplify, $2x + 8 = 130$. Subtract 8 from both sides, $2x = 122$. Divide both sides by 2; $x = 61$. The length of the rectangle is 61, and the width of the rectangle is $61 + 8$ or 69; $61 \times 69 = 4{,}209$. The second method of choice is to develop a system of equations using x and y. Let $x =$ the length of the rectangle and let $y =$ the width of the rectangle. Since the sum of the length and width of the rectangle is 130, we have the equation $x + y = 130$. The difference is 8, so we have the equation $x - y = 8$. If we add the two equations vertically, we get $2x = 122$. Divide both sides by 2: $x = 61$. The length of the rectangle is 61. Substitute 61 into either equation; $61 + y = 130$. Subtract 61 from both sides, giving you $y = 130 - 61 = 69$. To find the area of the rectangle, we use the formula *length* $\times$ *width* or $(61)(69) = 4{,}209$. If you chose **b**, you added 61 to 69 rather than multiplied. If you chose **c**, the length is 61 but the width was decreased by 8 to 53.

456. d. The volume of a sphere is found by using the formula $\frac{4}{3}\pi r^3$. Since the volume is 288π cm^3 and we are asked to find the radius, we will set up the following equation: $\frac{4}{3}\pi r^3 = 288\pi$. To solve for r, multiply both sides by 3; $4\pi r^3 = 864\pi$. Divide both sides by π; $4r^3 = 864$. Divide both sides by 4; $r^3 = 216$. Take the cube root of both sides; $r = 6$. If you chose **a**, the formula for volume of a sphere was incorrect; $\frac{1}{3}\pi r^3$ was used instead of $\frac{4}{3}\pi r^3$. If you chose **c**, near the end of calculations you mistakenly took the square root of 216 rather than the cube root.

457. d. $\angle ABE$ and $\angle CBD$ are vertical angles that are equal in measurement. Solve the following equation for x: $4x + 5 = 7x - 13$. Subtract $4x$ from both sides; $5 = 3x - 13$. Add 13 to both sides; $18 = 3x$. Divide both sides by 3; $6 = x$ or $x = 6$. To solve for $\angle ABE$ substitute $x = 6$ into the expression $4x + 5$ and simplify; $4(6) + 5$ equals $24 + 5$ or 29. $\angle ABE$ equals $29°$. If you chose **a**, you solved for $\angle ABC$ or $\angle EBD$. If you chose **b**, you assumed the angles were supplementary and set the sum of the two angles equal to 180.

458. **b.** If two angles are complementary, the sum of the measurement of the angles is 90°. $\angle 1$ is represented by x. $\angle 2$ is represented by $4x$. Solve the following equation for x: $x + 4x = 90$. Simplify; $5x = 90$. Divide both sides by 5; $x = 18$. The larger angle is $4x$ or $4(18)$, which equals 72°. If you chose **a**, the original equation was set equal to 180 rather than 90 and you solved for the smaller angle. If you chose **c**, the original equation was set equal to 180 rather than 90, and you solved for the larger angle. If you chose **d**, you solved the original equation correctly; however, you solved for the smaller of the two angles.

459. **d.** To find how far the wheel will travel, find the circumference of the wheel multiplied by 2. The formula for the circumference of the wheel is πd. Since the diameter of the wheel is 25 inches, the circumference of the wheel is 25π. Multiply this by 2, $(2)(25\pi)$ or 50π. Finally, substitute 3.14 for π; $50(3.14) = 157$ inches, the distance the wheel traveled in two turns. If you chose **a**, you used the formula for area of a circle rather than circumference. If you chose **b**, the distance traveled was one rotation, not two.

460. **a.** If two angles are supplementary, the sum of the measurement of the angles is 180°. $\angle 1$ is represented by x. $\angle 2$ is represented by $2x + 30$. Solve the following equation for x; $x + 2x + 30 = 180$. Simplify; $3x + 30 = 180$. Subtract 30 from both sides; $3x = 150$. Divide both sides by 3; $x = 50$. The larger angle is $2x + 30$ or $2(50) + 30$, which equals 130°. If you chose **b**, the equation was set equal to 90 rather than 180 and you solved for the smaller angle. If you chose **c**, x was solved for correctly; however, this was the smaller of the two angles. If you chose **d**, the original equation was set equal to 90 rather than 180, yet you continued to solve for the larger angle.

461. **d.** $\angle AED$ and $\angle BEC$ are vertical angles that are equal in measurement. Solve the following equation for x: $5x - 36 = 2x + 9$. Subtract $2x$ from both sides of the equation; $3x - 36 = 9$. Add 36 to both sides of the equation; $3x = 45$. Divide both sides by 3; $x = 15$. To solve for $\angle AED$ substitute $x = 15$ into the expression $2x + 9$ and simplify. $2(15) + 9$ equals 39. $\angle AED$ equals 39°. If you chose **a**, you solved for the wrong angle, either $\angle AEB$ or $\angle DEC$. If you chose **b**, you assumed the angles were supplementary and set the sum of the angles equal to 180°. If you chose **c**, it was the same error as choice **b**.

462. **a.** The sum of the measures of the angles of a triangle is 180°. Using this fact we can establish the following equation: $3x + 4x + 5x = 180$. Simplifying; $12x = 180$. Divide both sides by 12; $x = 15$. The largest angle is represented by $5x$. Therefore, $5x$, or $5(15)$, equals 75, the measure of the largest angle. If you chose **b**, the original equation was set equal to 90 rather than 180. If you chose **c**, this was the smallest angle within the triangle. If you chose **d**, this was the angle whose measurement lies between the smallest and largest angles.

463. **c.** The widest piece of mail will be equal to the length of the diagonal of the mailbox. The width, 4.5 in, will be a leg of the right triangle. The height, 5 in, will be another leg of the right triangle. We will solve for the hypotenuse, which is the diagonal of the mailbox, using the Pythagorean theorem; $a^2 + b^2 = c^2$ or $4.5^2 + 5^2 = c^2$. Solve for c, $20.25 + 25 = c^2$; $45.25 = c^2$; $c = 6.7$. If you chose **a**, you assigned the legs the values of 4.5 and 10; 10 is incorrect. If you chose **b**, you assigned the legs the values of 5 and 10. Again, 10 is incorrect.

464. **d.** To find the area of the cross section of pipe, we must find the area of the outer circle minus the area of the inner circle. To find the area of the outer circle, we will use the formula area = πr^2. The outer circle has a diameter of $4(3 + \frac{1}{2} + \frac{1}{2})$ and a radius of 2; therefore, the area = $\pi 2^2$ or 4π. The inner circle has a radius of 1.5; therefore, the area = $\pi(1.5)^2$ or 2.25π. The difference, $4\pi - 2.25\pi$ or 1.75π is the area of the cross section of pipe. If you chose **a**, you used the outer circle's radius of 3 and the inner circle's radius of $\frac{1}{2}$. If you chose **b** you used the outer circle's radius of $\frac{7}{2}$ and the inner circle's radius of 3. If you chose **c**, you used the outer radius of 4 and the inner radius of 3.

465. **a.** To find the volume of the pipe with a known cross section and length of 18 inches, simply multiply the area of the cross section times the length of the pipe. The area of the cross section obtained from the previous question was 1.75π in^2. The length is 18 inches. Therefore, the volume is 1.75 in^2 times 18 inches or 31.5π in^3. If you chose **b**, you multiplied choice **c** from the previous question by 18. If you chose **c**, you multiplied choice **a** from the previous question by 18. If you chose **d**, you multiplied choice **b** from the previous question by 18.

466. **c.** Sketching an illustration would be helpful for this problem. Observe
that point A is the starting point and point B is the ending point. After
sketching the four directions, we connect point A to point B. We can
add to the illustration the total distance traveled north as well as the
total distance traveled east. This forms a right triangle, given the
distance of both legs, with the hypotenuse to be solved. Using the
Pythagorean theorem, $a^2 + b^2 = c^2$, or $8^2 + 15^2 = c^2$; $64 + 225 = c^2$; $289 =$
c^2; $c = 17$. If you chose **a**, you mistakenly traveled 4 miles due east
instead of due west. If you chose **b**, you labeled the triangle incorrectly
by assigning 15 to the hypotenuse rather than a leg. If you chose **d**, you
solved the problem correctly but chose the wrong heading, northwest
instead of northeast.

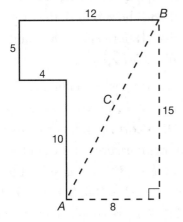

467. **b.** The area of the shaded region is the area of a rectangle, 22 by 12, minus
the area of a circle with a diameter of 12. The area of the rectangle is
$(22)(12) = 264$. The area of a circle with diameter 12 and a radius of 6, is
$\pi(6)^2 = 36\pi$. The area of the shaded region is $264 - 36\pi$. If you chose **a**,
the formula for area of a circle was incorrect, $\frac{1}{2}\pi r^2$. If you chose **c**, the
formula for area of a circle was incorrect, πd. If you chose **d**, this was the
reverse of choice **a**—area of the circle minus area of the rectangle.

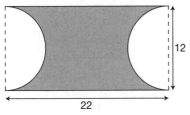

468. c. To find the area of the label, we will use the formula for the surface area of a cylinder, *area* = π*dh*, which excludes the top and bottom. Substituting *d* = 20 and *h* = 45, the area of the label is π(20)(45) or 900π cm². If you chose **a**, you used an incorrect formula for area, *area* = π*rh*. If you chose **b**, you used an incorrect formula for area, *area* = π*r²h*.

469. c. The sum of the measurement of ∠*AEB* and ∠*BEC* is 180°. Solve the following equation for *x*: 5*x* + 40 + *x* + 20 = 180. Simplify; 6*x* + 60 = 180. Subtract 60 from both sides; 6*x* = 120. Divide both sides by 6; *x* = 20. ∠*DEC* and ∠*AEB* are vertical angles that are equal in measurement. Therefore, if we find the measurement of ∠*AEB*, we also know the measure of ∠*DEC*. To solve for ∠*AEB*, substitute *x* = 20 into the equation 5*x* + 40 or 5(20) + 40, which equals 140°. ∠*DEC* is also 140°. If you chose **a**, you solved for ∠*BEC*. If you chose **b** or **d**, the original equation was set equal to 90 rather than 180. In choice **b**, you then solved for ∠*BEC*. In choice **d**, you solved for ∠*DEC*.

470. d. Two parallel lines cut by a transversal form corresponding angles that are congruent or equal in measurement. ∠*BAE* is corresponding to ∠*CFE*. Therefore ∠*CFE* = 46°. ∠*CDF* is corresponding to ∠*BEF*. Therefore, ∠*BEF* = 52°. The sum of the measures of the angles within a triangle is 180°. ∠*CFE* + ∠*BEF* + ∠*FGE* = 180°. Using substitution, 46 + 52 + ∠*FGE* = 180. Simplify; 98 + ∠*FGE* = 180. Subtract 98 from both sides; ∠*FGE* = 82°. ∠*FGE* and ∠*CGE* are supplementary angles. If two angles are supplementary, the sum of their measurements equals 180°. Therefore, ∠*FGE* + ∠*CGE* = 180. Using substitution, 82 + ∠*CGE* = 180. Subtract 82 from both sides; ∠*CGE* = 98°. If you chose **a**, you solved for ∠*CFE*. If you chose **b**, you solved for ∠*BEF*. If you chose **c**, you solved for ∠*FGE*.

471. b. To find the area of the shaded region, we must find the area of the circle minus the area of the rectangle. The formula for the area of a circle is πr^2. The radius is $\frac{1}{2}\overline{BC}$ or $\frac{1}{2}(10)$, which is 5. The area of the circle is $\pi(5^2)$ or 25π. The formula for the area of a rectangle is *length* × *width*. Using the fact that the rectangle is divided into two triangles with width of 6 and hypotenuse of 10, and using the Pythagorean theorem, we will find the length; $a^2 + b^2 = c^2$; $a^2 + 6^2 = 10^2$; $a^2 + 36 = 100$; $a^2 = 64$; $a = 8$. The area of the rectangle is *length* × *width* or 6 × 8 = 48. Finally, to answer the question, the area of the shaded region is the area of the circle – the area of the rectangle, or $25\pi - 48$. If you chose **a**, the error was in the use of the Pythagorean theorem, $6^2 + 10^2 = c^2$. If you chose **c**, the error was in finding the area of the rectangle. If you chose **d**, you used the wrong formula for area of a circle, πd^2.

472. b. The area of the shaded region is equal to the area of the square minus the area of the two semicircles. The area of the square is s^2 or 6^2, which equals 36. The area of the two semicircles is equal to the area of one circle. *Area* = πr^2 or $\pi(3)^2$ or 9π. Therefore, the area of the shaded region is $36 - 9\pi$. If you chose **a**, you calculated the area of the square incorrectly as 12. If you chose **c**, you used an incorrect formula for the area of two semicircles, $\frac{1}{2}\pi r^2$.

473. d. To solve for the length of the belt, begin with the distance from the center of each pulley, 3 ft, and multiply by 2; (3)(2) or 6 ft. Secondly, you need to know that the distance of two semicircles with the same radius is equivalent to the circumference of one circle. Therefore $C = \pi d$ or (12π) inches. Since the units are in feet, and not inches, convert (12π) inches to feet or (1π)ft. Now add these two values together, $(6 + 1\pi)$ft, to determine the length of the belt around the pulleys. If you chose **a** or **b**, you used an incorrect formula for circumference of a circle. Recall: *Circumference* = πd. If you chose **c**, you forgot to convert the unit from inches to feet.

474. **d.** To find the measure of an angle of any regular polygon, we use the formula $\frac{n-2}{n} \times 180$, where n is the number of sides. Using 15 as the value for n, $\frac{15-2}{15} \times 180 = \frac{13}{15} \times 180$ or 156. Alternatively, we compute the exterior angle of a regular polygon using the formula $\frac{360}{n}$, which in this case is $\frac{360}{15} = 24$. Since the angle we want and the exterior angle are supplementary, we again have the answer $180° - 24° = 156°$. If you chose **a**, you simply divided 360 (which is the sum of the exterior angles) by 15. If you chose **b**, you divided 180 by 15.

475. **c.** To find how many cubic yards of sand are in the pile, we must find the volume of the pile in cubic feet and convert the answer to cubic yards. The formula for volume of a cone is $V = \frac{1}{3}(height)(Area\ of\ the\ base)$. The area of the base is found by using the formula $Area = \pi r^2$. The area of the base of the sand pile is $\pi(16)^2$ or 803.84 ft^2. The height of the pile is 20 feet. The volume of the pile in cubic feet is $(803.84)(20)$ or 5,358.93 ft^3. To convert to cubic yards, divide 5,358.93 by 27 because 1 yard = 3 feet and 1 yd^3 means 1 yd $\times$ 1 yd $\times$ 1 yd which equals 3 ft $\times$ 3 ft $\times$ 3 ft or 27 ft^3. The answer is 198.5 yd^3. If you chose **a**, you did not convert to cubic yards. If you chose **b**, you converted incorrectly by dividing 5,358.93 by 9 rather than 27. If you chose **d**, the area of the base formula was incorrect. Area of a circle does not equal πd^2.

476. **b.** Observe that the octagon can be subdivided into 8 congruent triangles. Since each triangle has a base of 4 and a height of 7, the area of each traingle can be found using the formula, $area = \frac{1}{2}base \times height$. To find the area of the octagon, we will find the area of a triangle and multiply it by 8. The area of one triangle is $\frac{1}{2}(4)(7)$ or 14. Multiply this value times 8; $(14)(8) = 112$. This is the area of the octagon. If you chose **a**, you used an incorrect formula for area of a triangle. $Area = base \times height$ was used rather than $area = \frac{1}{2}\ base \times height$. If you chose **d**, you mistakenly divided the octagon into 6 triangles instead of 8 triangles.

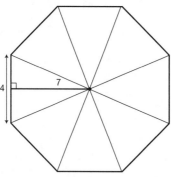

Another way to solve this problem is to use the formula for area of a regular polygon. That formula is *area* = $\frac{1}{2}Pa$, where P is the perimeter of the polygon and a is the apothem. If we know that the octagon is regular and each side is 4, that means the perimeter is 8 × 4 = 32. The apothem is the segment drawn from the center of the regular polygon and perpendicular to a side of the polygon; in this case it is 7. We substitute in our given values and get $\frac{1}{2}(32)(7) = 112$.

477. d. The sum of the measures of the angles of a quadrilateral is 360°. In the quadrilateral, three of the four angle measurements are known. They are 45° and two 90°angles. To find $\angle A$, subtract these three angles from 360°, or 360 – 90 – 90 – 45 = 135°. This is the measure of angle A. If you chose **a**, you assumed $\angle A$ and the 45° angle are complementary angles.

478. a. To find the total area of the shaded region, we must find the area of the rectangle minus the sum of the areas of all circles. The area of the rectangle is *length* × *width*. Since the rectangle is 4 circles long and 3 circles wide, and each circle has a diameter of 10 cm (radius of 5 cm × 2), the rectangle is 40 cm long and 30 cm wide; (40)(30) = 1,200 cm^2. The area of one circle is πr^2 or $\pi(5)^2 = 25\pi$. Multiply this value times 12, since we are finding the area of 12 circles, (12)(25)π = 300π. The difference is 1,200 – 300π cm^2, the area of the shaded region. If you chose **b**, the area of the rectangle was incorrectly calculated as (20)(15). If you chose **c**, you reversed the area of the circles minus the area of the rectangle. If you chose **d**, you reversed choice **b** as the area of the circles minus the area of the rectangle.

479. c. Referring to the illustration, $\angle NEB = 23°$ and $\angle DES = 48°$. Since $\angle NEB + \angle BED + \angle DES = 180$; using substitution, $23 + \angle BED + 48 = 180$. Simplify; $72 + \angle BED = 180$. Subtract 72 from both sides; $\angle BED = 109°$. If you chose **a**, you added $23 + 48$ to total 71. If you chose **b**, you assumed $\angle BED = \angle NEB$. If you chose **d**, you assumed $\angle BED = \angle DES$.

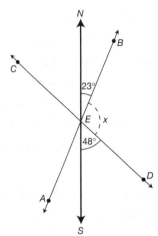

480. a. The measure of an angle of a regular polygon of n sides is $\frac{n-2}{n} \times 180$. Since a hexagon has 6 sides, to find the measure of $\angle ABC$, substitute $n = 6$ and simplify. The measure of $\angle ABC$ is $\frac{6-2}{6} \times 180$ or $120°$. If you chose **b**, you assumed a hexagon has 8 sides. If you chose **c**, you assumed a hexagon has 5 sides. If you chose **d**, you assumed a hexagon has 10 sides.

481. d. The volume of a box is found by multiplying *length* × *length* × *length* or $l \times l \times l = l^3$. If the length is doubled, the new volume is $(2l) \times (2l) \times (2l)$ or $8(l^3)$. When we compare the two expressions, we can see that the difference is a factor of 8. Therefore, the volume has been increased by a factor of 8.

482. a. The formula for finding the circumference of a circle is πd. If the radius is tripled, the diameter is also tripled. The new circumference is $\pi 3d$. Compare this expression to the original formula; with a factor of 3, the circumference is multiplied by 3.

483. a. The formula for the surface area of a sphere is $4\pi r^2$. If the diameter is doubled, this implies that the radius is also doubled. The formula then becomes $4\pi (2r)^2$. Simplifying this expression, $4\pi(4r^2)$ equals $16\pi r^2$. Compare $4\pi r^2$ to $16\pi r^2$; $16\pi r^2$ is 4 times greater than $4\pi r^2$. Therefore, the surface area is four times as great.

484. b. If the diameter of a sphere is tripled, the radius is also tripled. The formula for the volume of a sphere is $\frac{4}{3}\pi r^3$. If the radius is tripled, $volume = \frac{4}{3}\pi(3r)^3$ which equals $\frac{4}{3}\pi(27r^3)$ or $\frac{4}{3}(27)\pi r^3$. Compare this equation for volume with the original formula; with a factor of 27, the volume is now 27 times as great.

485. b. The formula for the volume of a cone is $\frac{1}{3}\pi r^2 h$. If the radius is doubled, then $volume = \frac{1}{3}\pi(2r)^2 h$ or $\frac{1}{3}\pi 4r^2 h$. Compare this expression to the original formula; with a factor of 4, the volume is multiplied by 4.

486. a. The formula for the volume of a cone is $\frac{1}{3}\pi r^2 h$. If the radius is halved, the new formula is $\frac{1}{3}\pi(\frac{1}{2}r)^2 h$ or $\frac{1}{3}\pi(\frac{1}{4})r^2 h$. Compare this expression to the original formula; with a factor of $\frac{1}{4}$, the volume is multiplied by $\frac{1}{4}$.

487. b. The volume of a right cylinder is $\pi r^2 h$. If the radius is doubled and the height halved, the new volume is $\pi(2r)^2(\frac{1}{2}h)$ or $\pi 4r^2(\frac{1}{2}h)$ or $2\pi r^2 h$. Compare this expression to the original formula; with a factor of 2, the volume is multiplied by 2.

488. a. The formula for finding the volume of a right cylinder is $volume = \pi r^2 h$. If the radius is doubled and the height is tripled, the formula has changed to $\pi(2r)^2(3h)$. Simplified, $\pi 4r^2 3h$ or $\pi 12r^2 h$. Compare this expression to the original formula; with a factor or 12, the volume is now multiplied by 12.

489. c. The measure of an angle of a regular polygon of n sides is $\frac{n-2}{n} \times 180$. Since each angle measures 150°, we will solve for n, the number of sides. Using the formula $150 = \frac{n-2}{n} \times 180$, solve for n. Multiply both sides by n, $150n = (n-2)180$. Distribute by 180, $150n = 180n - 360$. Subtract $180n$ from both sides, $-30n = -360$. Divide both sides by -30, $n = 12$. The polygon has 12 sides. Alternately, if each angle is 150°, each exterior angle is $180° - 150° = 30°$. The measure of an exterior angle of a rectangular polygon of n sides is $\frac{360}{n}$. We now solve for n by solving $\frac{360}{n} = 30$. Multiplying both sides by n, $360 = 30n$. Dividing both sides by 30, $n = 12$.

490. d. This problem requires two steps. First, find the diagonal of the base of the box. Second, using this value, find the length of the diagonal $\overline{AB}$. To find the diagonal of the base, use 30 cm as a leg of a right triangle, 8 cm as the second leg, and solve for the hypotenuse. Using the Pythagorean theorem, $30^2 + 8^2 = c^2$; $900 + 64 = c^2$; $964 = c^2$; $c = 31.05$. Now consider this newly obtained value as a leg of a right triangle, 12 cm as the second leg, and solve for the hypotenuse, $\overline{AB}$; $31.05^2 + 12^2 = \overline{AB}^2$; $964 + 144 = \overline{AB}^2$; $1,108 = \overline{AB}^2$. $\overline{AB} = 33.3$. If you chose **a**, you used 30 and 12 as the measurements of the legs. If you chose **b**, you solved the first triangle correctly; however, you used 8 as the measure of one leg of the second triangle, which is incorrect.

491. c. To find the area of the shaded region, we must find $\frac{1}{2}$ the area of the circle with diameter AC, minus $\frac{1}{2}$ the area of the circle with diameter BC, plus $\frac{1}{2}$ the area of the circle with diameter AB. To find $\frac{1}{2}$ the area of the circle with diameter AC, we use the formula $area = \frac{1}{2}\pi r^2$. Since the diameter is 6, the radius is 3; therefore, the area is $\frac{1}{2}\pi 3^2$ or 4.5π. To find $\frac{1}{2}$ the area of the circle with diameter BC, we again use the formula $area = \frac{1}{2}\pi r^2$. Since the diameter is 4, the radius is 2; therefore the area is $\frac{1}{2}\pi 2^2$ or 2π. To find $\frac{1}{2}$ the area of the circle with diameter AB we use the formula $area = \frac{1}{2}\pi r^2$. Since the diameter is 2, the radius is 1; therefore the area is $\frac{1}{2}\pi$. Finally, $4.5\pi - 2\pi + .5\pi = 3\pi$, the area of the shaded region. If you chose **a** or **b**, in the calculations you mistakenly used πd^2 as the area formula rather than πr^2.

492. a. This problem has three parts. First, we must find the diameter of the existing tower. Secondly, we will increase the diameter by 16 meters for the purpose of the fence. Finally, we will find the circumference using this new diameter. This will be the length of the fence. The formula for circumference of a circle is πd. This formula, along with the fact that the tower has a circumference of 40 meters, gives us the following formula: $40 = \pi d$. To solve for d, the diameter, divide both sides by π. $D = \frac{40}{\pi}$ the diameter of the existing tower. Now increase the diameter by 16 meters; $\frac{40}{\pi} + 16$ to get the diameter of the fenced in section. Finally, use this value for d in the equation πd or $\pi(\frac{40}{\pi} + 16)$ meters. Simplify by distributing π through the expression; (40 + 16π) meters. This is the length of the security fence. If you chose **b**, you added 8 to the circumference of the tower rather than 16. If you chose **c**, you merely added 8 to the circumference of the tower.

493. a. Using the illustration, ∠2 = ∠*a*. ∠2 and ∠*a* are vertical angles. ∠1 and ∠*a* are supplementary, since ∠*c* + ∠*d* + ∠1 + ∠*a* = 360° (the total number of degrees in a quadrilateral), 90 + 90 + ∠1 + ∠*a* = 360. Simplifying, 180 + ∠1 + ∠*a* = 360. Subtract 180 from both sides; ∠1 + ∠*a* = 180. Since ∠*a* = ∠2, using substitution, ∠1 + ∠2 = 180. Using similar logic, ∠4 = ∠*b*. ∠4 and ∠*b* are vertical angles. ∠3 and ∠*b* are supplementary. ∠*e* + ∠*f* + ∠*b* + ∠3 = 360 or 90 + 90 + ∠*b* + ∠3 = 360. Simplifying, 180 + ∠*b* + ∠3 = 360. Subtract 180 from both sides, ∠*b* + ∠3 = 180. Since ∠*b* = ∠4, using substitution, ∠3 + ∠4 = 180. Finally, adding ∠1 + ∠2 = 180 to ∠3 + ∠4 = 180, we can conclude ∠1 + ∠2 + ∠3 + ∠4 = 360.

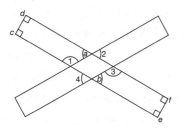

494. a. To find the volume of the hollowed solid, we must find the volume of the original cone minus the volume of the smaller cone sliced from the original cone minus the volume of the cylindrical hole. The volume of the original cone is found by using the formula $V = \frac{1}{3}\pi r^2 h$. Using the values $r = 9$ and $h = 40$, substitute and simplify to find the volume = $\frac{1}{3}\pi(9)^2(40)$ or $1,080\pi$ cm³. The volume of the smaller cone is found by using the formula $V = \frac{1}{3}\pi r^2 h$. Using the values $r = 3$ and $h = 19$, substitute and simplify to find the volume = $\frac{1}{3}\pi(3)^2(19)$ or 57π cm³. The volume of the cylinder is found by using the formula $V = \pi r^2 h$. Using the values $r = 3$ and $h = 21$, substitute and simplify to find the volume = $\pi(3)^2(21)$ or 189π cm³. Finally, calculate the volume of the hollow solid; $1,080\pi - 57\pi - 189\pi$ or 834π cm³. If you chose **b**, you used an incorrect formula for the volume of a cone, $V = \pi r^2 h$. If you chose **c**, you subtracted the volume of the large cone minus the volume of the cylinder. If you chose **d**, you added the volumes of all three sections.

495. c. To find the volume of the object, we must find the volume of the water that is displaced after the object is inserted. Since the container is 5 cm wide and 15 cm long, and the water rises 2.3 cm after the object is inserted, the volume of the displaced water can be found by multiplying length by width by depth: $(5)(15)(2.3) = 172.5$ cm^3.

496. a. To find how many cubic yards of concrete are needed to construct the wall, we must determine the volume of the wall. The volume of the wall is calculated by finding the surface area of the end and multiplying it by the length of the wall, 120 ft. The surface area of the wall is found by dividing it into three regions, calculating each region's area, and adding them together. The regions are labeled A, B, and C. To find the area of region A, multiply the length (3) times the height (10) for an area of 30 ft^2. To find the area of region B, multiply the length (5) times the height (3) for an area of 15 ft^2. To find the area of region C, multiply $\frac{1}{2}$ times the base (2) times the height (4) for an area of 4 ft^2. The surface area of the end is 30 ft^2 + 15 ft^2 + 4 ft^2 or 49 ft^2. Multiply 49 ft^2 by the length of the wall 120 ft; 5,880 ft^3 is the volume of the wall. The question, however, asks for the answer in cubic yards. To convert cubic feet to cubic yards, divide 5,880 ft^3 by 27 ft^3, the number of cubic feet in one cubic yard, which equals 217.8 yd^3. If you chose **b**, you did not convert to yd^3. If you chose **c**, the conversion to cubic yards was incorrect. You divided 5,880 by 9 rather than 27. If you chose **d**, you found the area of the end of the wall and not the volume of the wall.

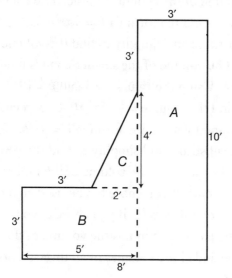

497. b. To find the volume of the sphere we must find the volume of the outer sphere minus the volume of the inner sphere. The formula for volume of a sphere is $\frac{4}{3}\pi r^3$. The volume of the outer sphere is $\frac{4}{3}\pi(120)^3$. Here the radius is 10 feet (half the diameter) multiplied by 12 (converted to inches), or 120 inches. The volume equals 7,234,560 in^3. The volume of the inner sphere is $\frac{4}{3}\pi(119)^3$ or 7,055,199. (This is rounded to the nearest integer value.) The difference of the volumes is 7,234,560 − 7,055,199 or 179,361 in^3. This answer is in cubic inches, and the question is asking for cubic feet. Since one cubic foot equals 1,728 cubic inches, we simply divide 179,361 by 1,728, which equals 104, rounded to the nearest integer value. As an alternative to changing units to inches only to have to change them back into feet again, keep units in feet. The radius of the outer sphere is 10 feet and the radius of the inner sphere is one inch less than 10 feet, which is 9 and $\frac{11}{12}$ feet, or 9.917 feet. Use the formula for volume of a sphere: $\frac{4\pi r^3}{3}$ and find the difference in the volumes. If you chose **a**, you used an incorrect formula for the volume of a sphere, $V = \pi r^3$. If you chose **c**, you also used an incorrect formula for the volume of a sphere, $V = \frac{1}{3}\pi r^3$. If you chose **d**, you found the correct answer in cubic inches; however, your conversion to cubic feet was incorrect.

498. c. To solve this problem, we must find the volume of the sharpened tip and add this to the volume of the remaining lead that has a cylindrical shape. To find the volume of the sharpened point, we will use the formula for finding the volume of a cone, $\frac{1}{3}\pi r^2 h$. Using the values $r =$ .0625 (half the diameter) and $h = .25$, the volume $= \frac{1}{3}\pi(.0625)^2(.25)$ or .002 in^3. To find the volume of the remaining lead, we will use the formula for finding the volume of a cylinder, $\pi r^2 h$. Using the values $r = .0625$ and $h = 5$, the volume $= \pi(.0625)^2(5)$ or .0613. The sum is .001 + .0613 or .0623 in^3, the volume of the lead. If you chose **a**, this is the volume of the lead without the sharpened tip. If you chose **b**, you subtracted the volumes calculated.

499. b. To find the volume of the hollowed solid, we must find the volume of the cube minus the volume of the cylinder. The volume of the cube is found by multiplying *length × width × height* or (5)(5)(5) equals 125 in³. The value of the cylinder is found by using the formula $\pi r^2 h$. In this question, the radius of the cylinder is 2 and the height is 5. Therefore, the volume is $\pi(2)^2(5)$ or 20π. The volume of the hollowed solid is 125 − 20π. If you chose **a**, you made an error in the formula of a cylinder, using $\pi d^2 h$ rather than $\pi r^2 h$. If you chose **c**, this was choice **a** reversed. This is the volume of the cylinder minus the volume of the cube. If you chose **d**, you found the reverse of choice **b**.

500. b. Refer to the diagram to find the area of the shaded region. One method is to enclose the figure into a rectangle, and subtract the area of the unwanted regions from the area of the rectangle. The unwanted regions have been labeled *A* through *F*. The area of region *A* is (15)(4) = 60. The area of region *B* is (5)(10) = 50. The area of region *C* is (20)(5) = 100. The area of region *D* is (17)(3) = 51. The area of region *E* is (20)(5) = 100. The area of region *F* is (10)(5) = 50. The area of the rectangle is (23)(43) = 989. The area of the shaded region is 989 − 60 − 50 − 100 − 51 − 100 − 50 = 578. If you chose **a**, **c** or **d**, you omitted one or more of the regions *A* through *F*.

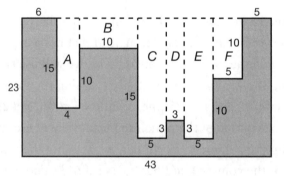

501. **d.** The shape formed by the paths of the two arrows and the radius of the bull's eye is a right triangle. The radius of the bull's eye is one leg and the distance the second arrow traveled is the second leg. The distance the first arrow traveled is the hypotenuse. To find the distance the first arrow traveled, use the Pythagorean theorem where 2 meters (half the diameter of the target) and 20 meters are the lengths of the legs and the length of the hypotenuse is missing. Therefore, $a^2 + b^2 = c^2$ and $a = 2$ and $b = 20$, so $2^2 + 20^2 = c^2$. Simplify: $4 + 400 = c^2$. Simplify: $404 = c^2$. Find the square root of both sides: $20.1 = c$. So the first arrow traveled about 20.1 meters. If you chose **c**, you added the two lengths together without squaring. If you chose **b**, you added Kim's distance from the target to the diameter of the target. If you chose **a**, you let 20 meters be the hypotenuse of the right triangle instead of a leg and you used the radius of the target.

Special Offer from LearningExpress!

Let LearningExpress help you acquire practical, essential math word problems skills FAST

Go to the LearningExpress Practice Center at www.LearningExpress FreeOffer.com, an interactive online resource available exclusively for LearningExpress customers.

Now that you've purchased LearningExpress's *501 Math Word Problems, 2nd Edition*, you have **FREE** access to:

- **40 practice questions** covering different types of math word problems from fractions to decimals to percents to algebra and geometry
- **Immediate scoring** and **detailed answer explanations**
- Benchmark your skills and focus your study with our **customized diagnostic report**

Follow the simple instructions on the scratch card in your copy of *501 Math Word Problems, 2nd Edition*. Use your individualized access code found on the scratch card and go to www.LearningExpressFreeOffer.com to sign in. Improve your math skills right away!

Once you've logged on, use the spaces below to write in your access code and newly created password for easy reference:

Access Code: _____   Password: _____